NATHANAEL-ISRAEL ISRAEL, PhD

How Baby Universe Was Born

OTHER BOOKS BY NATHANAEL-ISRAEL ISRAEL

Get them at your local bookstore, or online (e.g. on Amazon, Science180.com/books)

Turbulent Origin of the Universe
There is Only One Scientific, Simple, Safe, Trustworthy, Unexpensive, Brave, Practical, Nonconformist, Universal, Verifiable Formula that Accurately Decodes the Universe Formation … But You Are Not Using It

Reconciling Science and Creation Accurately
What Science Accurately Teaches about Creation and God's Existence that Atheists, Freethinkers, and even Most Christians Ignore … And How to Demonstrate it Without Taking Sides Between Rationality and Faith

Turbulent Origin of Chemical Particles
Why You Don't Have to Embrace Evolution, Big Bang, or Deny God to Scientifically Prove the Formation of All Chemical Particles

Origin of the Spiritual World
Top Secrets about the Origin of Everything in the Universe that Some Elites Have Hidden from You for Thousands of Years

From Science to Bible's Conclusions
How Decoding the Universe-Origin by Properly Revisiting Scientific Data—That Top Scientists Collected but Wrongly Analyzed—Bizarrely led to the 3500 Years Old Biblical Account of Creation

Turbulent Origin of Life
Why You Don't Have to Embrace Evolutionism or Check Your Brain at the Door in the Name of Faith or Science to Accurately Decrypt the Origin of Life Using the Historic Formula of the Universe Formation

How God Created Baby Universe
What Children Must Scientifically Learn Early about the Universe Formation to Avoid Dangerously Abandoning God Later in Life Just Like Most College Students Who Embrace Evolution and Big Bang That Deny Biblical Creation

Science180 Accurate Scientific Proof of God
Can We Scientifically Explain the Formation of the Universe Through Natural Processes Without Evoking Evolution and Big Bang?

Mathematical Proof of God's Existence at the Intersection of Science and Faith.
The Scientifically Verifiable Cosmological Theory that Challenges the Big Bang Theory at the Crossroads of Reason and Religion THEY Want You to Ignore

Boldest Scientific Formula of God and Creation
Live Better by Quickly Mastering the Only Scientific Theory that Evolution, Big Bang, AI, and some Religions Desperately Ache YOU Ignore
More books written by Nathanael-Israel Israel can be found at Israel120.com/books

NATHANAEL-ISRAEL ISRAEL, PhD

Founder of Science180 and of Science180 Academy
Discoverer of the "Universe Turbulent Origin Formula"
Discoverer of the "Life Turbulent Origin Formula"
www.Science180.com

How Baby Universe Was Born

How to Scientifically Talk to Children about the Universe Formation and They will Know Forever How to Correctly Test the Intersection of Science and Faith

Science180

Augusta
United States of America
www.Science180Publishing.com

How Baby Universe Was Born
How to Scientifically Talk to Children about the Universe Formation and They will Know
Forever How to Correctly Test the Intersection of Science and Faith

First edition: October 2025

Published by Science180
Augusta (USA)
www.Science180Publishing.com

Book Cover and Illustrations by Nathanael-Israel Israel

ISBN: 979-8-9932150-8-2

Library of Congress Control Number: 2025920907

More books by the same author can be found at Israel120.com and Science180.com

For information about special discounts available for bulk purchases, please visit
Science180.com/discount for more details.

Science180 can bring authors including Dr. Nathanael-Israel Israel to your live or recorded
events. For more information or to book an event, please visit Science180.com/speaking

For any questions, please visit Science180.com/contact

To publish your book(s) with Science180 Publishing, go to Science180Publishing.com

To interview the author of this book, visit Israel120.com/interview
To donate, please visit Israel120.com/donate or Science180.com/donate.

Printed in the United States of America.

CONTENT

SECTION 1: INTRODUCTION

Science180: The All-In-One Proven & Uncomplicated Universe-Origin and Life-Origin Formula

1. WHO AM I, AND WHY IS THIS A FANTASTIC BOOK?

- How was the universe formed?
- How were the Sun and all the stars in the universe formed?
- How were the Earth and all the other planets in the universe formed?
- How were the Moon and the other satellites in the universe formed?
- Is there any book that presents a nice story of the formation of the universe that both children and their parents can read and enjoy together as a family?
- How can children know for sure which story of the origin of the universe is trustworthy and correct?
- Is there any story on the origin of the universe that can prepare children for a better tomorrow?
- How can children correct wrong stories about the origin of the universe so they can save time, money, and improve their lives?
- How can children be trained to properly answer deep questions about the formation of the Earth, Moon, and the Sun that even some highly educated people ignore?
- Can children use mathematics and science to test whether God created the universe as the Bible says, or whether billions of years of evolutionary processes formed it?
- How can children use science to test the existence of God?
- How can we explain all of these difficult things to children in simple language that they can easily understand and enjoy?

If you are interested in finding the correct answer to any of these questions and others related to the origin of the universe and everything in it, including the planets, stars, satellites, and even life, then you have found the perfect book.

Welcome to *How Baby Universe Was Born*", a well-known book that all children and the people dear to them enjoy. It is about how the entire universe was formed. As you read this book, I am sure you will learn a lot about how the universe began. Before I start telling the story, please let me tell you a little bit about myself.

My name is Josephine Israel. I have a sister (Joelle-Major Sophia-Arielle Israel) and a brother (Joshua-Enoch Michael-Uriel Israel). I am 10 years old. My little sister

SECTION 1: INTRODUCTION

is 9, and my little brother is 7. We are all in elementary school. Our dad is a scientist, a writer, and a businessman. He has spent many years studying and researching how the universe was formed. He discovered many great things about the origin of the universe, which means how the universe was formed. I am not surprised that people tell my dad he is the #1 international authority who truly helps people properly unlock the secrets of the turbulence that shaped the universe. He is acknowledged as the world's most accurate universe-origin scientist and the world's most trusted expert for properly decoding the formation of the universe, life, and chemicals. He is the "Creator of Science180 Academy" (www.Science180Academy.com) and the "Creator of the Universe Turbulent Origin Formula". Although people call him by many other names, my siblings and I are glad just to call him Daddy.

In 2025, he published some of his discoveries in 9 books, including:

- "Turbulent Origin of the Universe" (written for scientists: people who study science)
- "Turbulent Origin of Chemical Particles" (written for chemists: people who study chemicals)
- "Turbulent Origin of Life" (written for biologists (people who study biology or life science) and for anyone else who wants to understand how life was formed)
- And 5 other books besides the one you are reading now

But none of his initial books were written for children like us. Therefore, Daddy decided to share with us what he has discovered about the formation of the universe in a language that children can understand. Because we were happy with the lessons, we wanted to share them with the whole world. In other words, our Daddy thought it would be even better to write a book about how the universe was formed so that children our age around the globe can understand and enjoy it.

In this book, I will be leading my little sister and brother to share with you great information about how the entire universe came to be. Later, we will do some math to calculate how long it took for the Earth, the Moon, and the Sun to be formed. Then we will see whether that story really matches what the Bible or other books say.

The first children's book my dad wrote about the origin of the universe was for people, including Christians who believe in God, as well as anyone who wants to know how God created the universe. Hence, he named that book "*How God Created Baby Universe*". That book is divided into 2 parts. The first part dealt with pure science, and the second addressed many questions that smart children and their parents often ask about God.

Two years after writing that book, my dad realized that some people who don't believe in God but are also really interested in the real origin of the universe may not read it because it talks about God. Therefore, to help unbelievers or God-deniers rationally know at least how his discovery points to an original story of the formation of the universe that also scientifically tests the existence of God, Dad decided to remove the second part of his initial children's book and make the title

Science180: The New Physics and Life-Origin Science that Will Revolutionize Science Forever

and the content fit those who deny God. That is how this scientific book, *"How Baby Universe Was Born"*, came into existence in early 2025. It is for children ages 7-12 who don't or whose parents don't believe in God, but who want to dig into the real beginning and formation of the universe using pure science.

Everybody wants to know about the universe, and I am glad my dad has found a way to share his cool discovery with curious children like us. This book definitely presents the origin story in a style that will challenge you to think creatively. I am sure you will like it. Without waiting any longer, just relax, let's go!

2. VERY IMPORTANT QUESTIONS ABOUT THE UNIVERSE'S FORMATION

After Dad decided to write the book for children like us, he ensured we would like it. To be sure he would answer all of our questions, Dad gave us a week to come up with the questions we had about how the universe was formed. Many very important questions quickly came to mind, but we were not able to answer them properly, and we were not even sure our answers were correct. By the way, whenever I use the word "we," please know I am talking about my little brother, my sister, and me.

To the surprise of our dad, within a few days, we came up with more than 100 great and very big questions that we wanted to understand:
1. How was Baby Earth born?
2. Who was Baby Earth's mother?
3. How old is the Earth?
4. How was the Moon formed?
5. How were all the planets formed?
6. How were the stars formed?
7. Why does the Moon turn around the Earth?
8. Why does the Earth turn around the Sun?
9. Why does the Sun give light while the Earth does not?
10. How were the seas and the oceans made?
11. How was the sky made?
12. Why are the planets different?
13. Why are some planets gaseous, others solid, and some icy?
14. Why do planets have different colors?
15. How were plants made?
16. How were animals and people made?
17. How many creatures are there in the universe?
18. Why do birds fly, but people cannot fly?
19. How was the air we breathe made, and where do clouds come from?
20. Why is there light and darkness, and where do they come from?
21. How can we use science to test whether God really created the universe as the Bible says? etc.

These questions are very important, and we need to know their answers so we can be smarter. As we presented these questions to Dad, he was happy because he felt like they were really HUGE problems. Although these questions are very hard for children our age, Dad said he will do his best to help us overcome any obstacles that may try to prevent us from understanding them. If you have ever asked any of these questions or are interested in learning about them, you are in the right place. With this amazing origin of the universe book that your whole family will like and enjoy together, you will:

Science180: The New Physics and Life-Origin Science that Will Revolutionize Science Forever

- Have peace of mind that you will get accurate, fit, and easy-to-understand universe-origin information that will transform your life
- Boost your confidence in detecting, confronting, and avoiding wrong theories by knowing the facts and real processes involved in the formation of the universe
- Know how to easily sort out great questions using strategies that tap into deep secrets that even highly educated people ignore
- Clearly understand how to mathematically know without a doubt whether God created the universe as the Bible says, or whether billions of years of evolutionary processes formed it

Throughout my writing, wherever you see "universe-origin," please know that I meant "origin of the universe" or "the origin of the universe." Likewise, wherever you see "life-origin," please understand that I meant "origin of life" or "the origin of life." In the same manner, wherever I mention "chemicals-origin," please know that I am referring to "origin of chemicals" or "the origin of chemicals." Moreover, I will be posting some interesting kids' content online, and you can find it at www.Science180.com/childrensecular.

In the rest of this book, I will share with you what my siblings and I have learned from Dad about how the universe was formed. Do you want to get started with some fun stuff that will make you laugh a lot? If yes, let's go!

SECTION 2: HOW THE GALAXIES, THE PLANETS, THE MOON, AND THE SUN WERE FORMED

3. IMPORTANT THINGS IN THE SOLAR SYSTEM

When I look in the sky, I see the stars, the moon, the clouds, some planets, and the Sun. I learned that all of those stars, planets, and moons are moving in space without anything holding them. I learned that each of them was created on a specific date. As I was curious, I asked Daddy, "How was the Earth made?"

Daddy told me that to better explain how the Earth was formed, he first needed to tell me certain things about the Earth.

The Solar System (see Figure 1) is the name by which the Sun and everything that orbits the Sun are called. In other words, the Solar System is a group or family of celestial bodies made of the Sun and the bodies that orbit the Sun. The Earth is one of the many planets in the Solar System. To explain how the Earth was made, Daddy said he also needed to explain how the W-H-O-L-E Solar System was made.

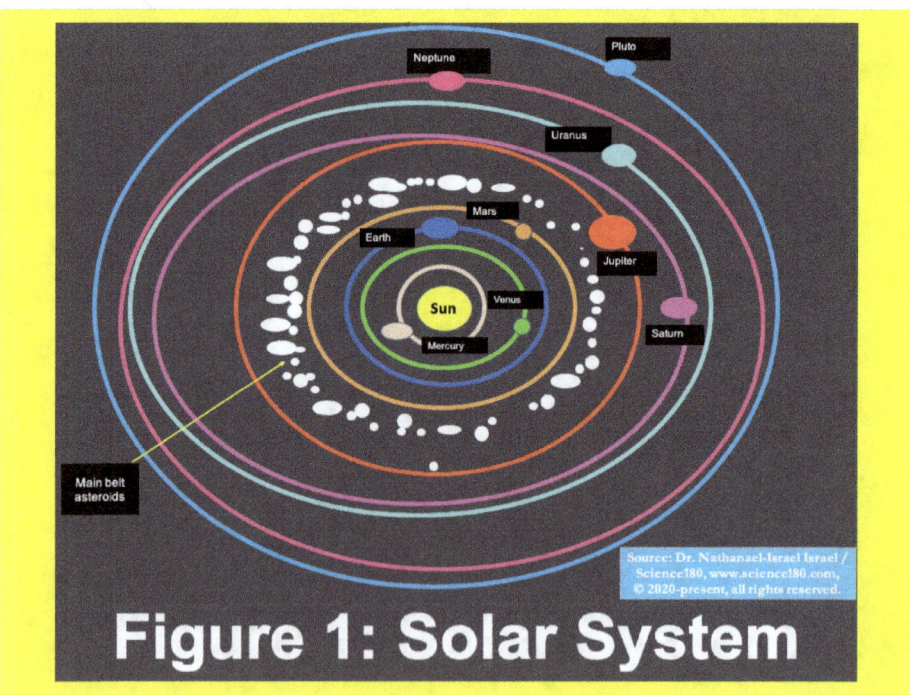

Figure 1: Solar System

For thousands of years, people have tried to explain how the Solar System was made, and they have come up with many ideas. I will not go over all of those ideas with you because you may not understand them, as they involve a lot of math, a lot of thinking, and a lot of guessing, which is hard to believe. But I will tell you what I learned from Dad, who spent many years to figure out how the WHOLE universe was formed. Let's first look at certain things in the Solar System.

Nathanael-Israel Israel: Known as the World's Most Accurate Universe-Origin Mathematician

SECTION 2: HOW THE GALAXIES, THE PLANETS, THE MOON, AND THE SUN WERE FORMED

The Sun, the Earth, the Moon, and all of the other planets in the Solar System are called celestial bodies. In other words, a celestial body is a big natural object like a planet or a star in the sky.

The Sun is the biggest celestial body in the Solar System. The Sun is almost 1000 times as big as Earth. The Earth is a planet. The other planets in the Solar System are Mercury, Venus, Mars, Jupiter, Saturn, Uranus, and Neptune. For a long time, Pluto was considered a planet, but for a few years, some people haven't considered it a planet anymore. However, in many books, you will see Pluto still considered a planet. Old people who learned the names of the planets a long time ago will still tell you that Pluto is a planet. Therefore, forgive me or anybody else if sometimes I or they also consider Pluto a planet.

The Sun is located almost in the middle of the Solar System. All of the other celestial bodies in the Solar System are moving around the Sun, and each of them is located at a specific distance. Some are very close to the Sun, and others are very far away, but all of them are located millions of miles from the Sun and from one another. Mercury is the closest planet to the Sun.

Certain planets have other celestial bodies turning around them. For example, the Moon turns around the Earth (see Figure 2).

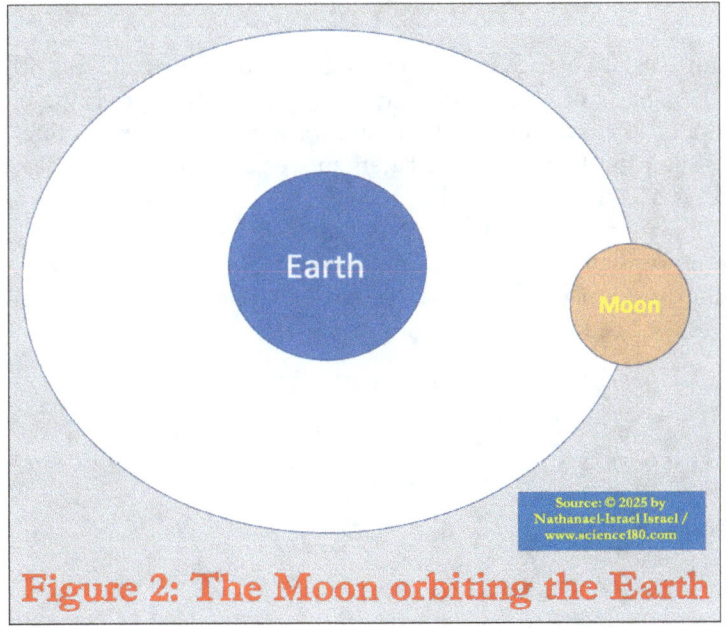

Earth

Moon

Figure 2: The Moon orbiting the Earth

A celestial body that orbits a planet is called a satellite. Some planets do not have even a single satellite. For instance, Mercury and Venus do not have a satellite.

Some celestial bodies turn around the Sun but are not satellites. Most of them are called asteroids, and all of them are smaller than the planets. For instance, between Mars and Jupiter is the main asteroid belt, which contains millions of asteroids (see Figure 3).

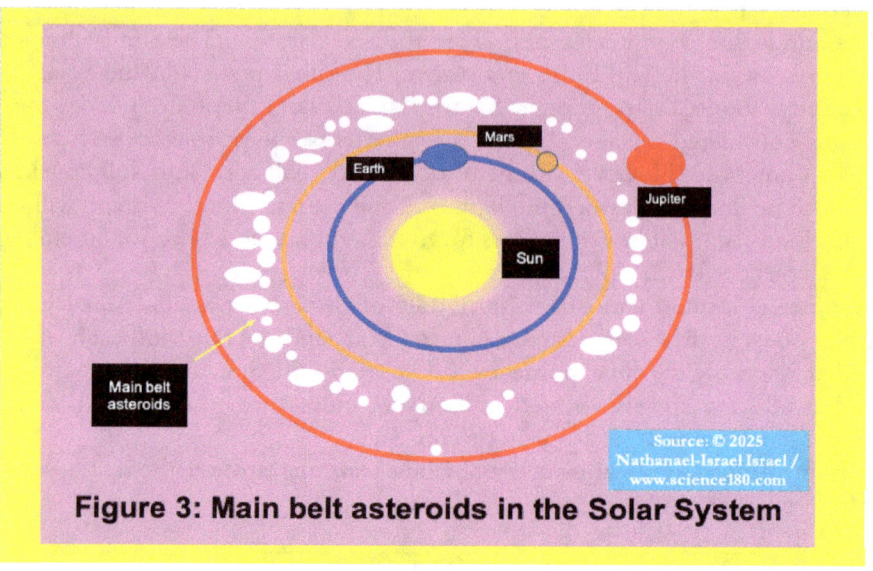

Figure 3: Main belt asteroids in the Solar System

Each body in the Solar System is different. Some are small, and others are big. The Sun is the biggest celestial body in the Solar System. Jupiter is the biggest planet moving around the Sun. Mars and Pluto are the smallest planets. Jupiter is about 10 times bigger than the Earth. Some planets move fast, and some move slowly. The fastest planet is Mercury, and the slowest planet is Pluto. For those who think Pluto is no longer a planet, we can say that Neptune is the slowest planet. Neptune is the planet just before Pluto. Daddy told me that from Mercury all the way to Neptune and Pluto, the speed of the planets decreases, meaning it slows as the distance from the Sun increases. See Figure 4 for more details.

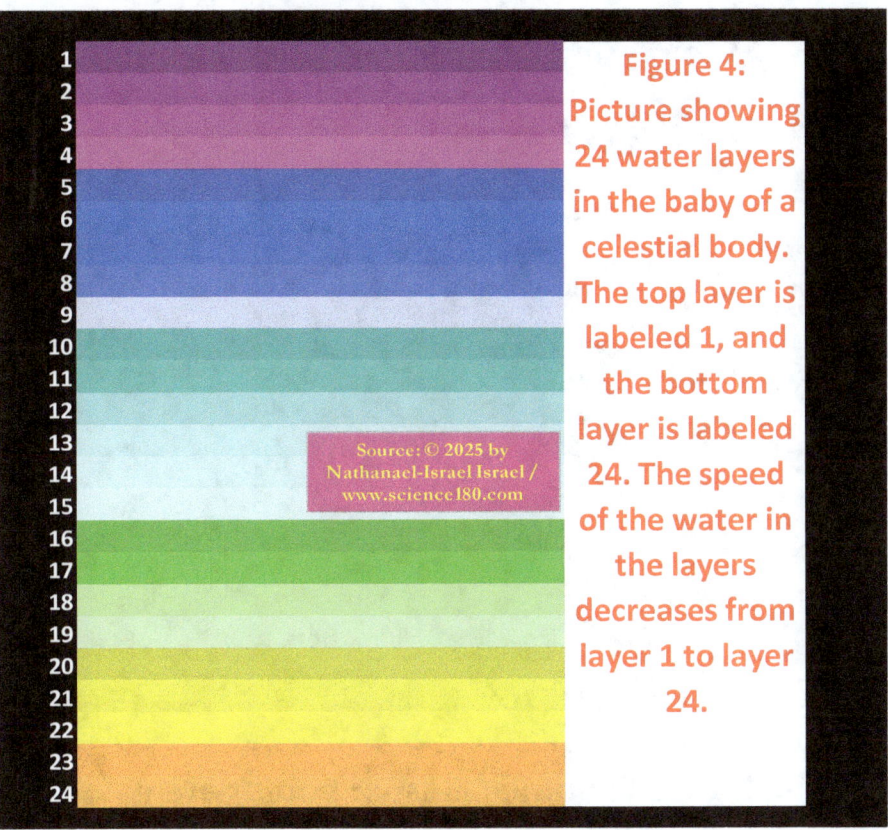

Figure 4: Picture showing 24 water layers in the baby of a celestial body. The top layer is labeled 1, and the bottom layer is labeled 24. The speed of the water in the layers decreases from layer 1 to layer 24.

Source: © 2025 by Nathanael-Israel Israel / www.science180.com

The celestial bodies in the Solar System are made of different materials. Some are very hot, and others are very cold. For example, the Sun is the HOTTEST celestial body in the Solar System, and the farther we get from the Sun, the colder things get. In other words, the celestial bodies that are FAR away from the Sun are colder than those close to the Sun. Some celestial bodies in the Solar System are solid, some are ice, and some are gas. By the way, the gas I'm talking about here is not the gas we put in our cars or the air we release when we use the bathroom. Thank goodness, or else those celestial bodies made up of gas would be STINKY! The Earth is a very solid planet. Jupiter is an example of a gas giant. Uranus is a planet that is filled with ice, like the ice we get from the freezer on a hot summer day (Figure 5).

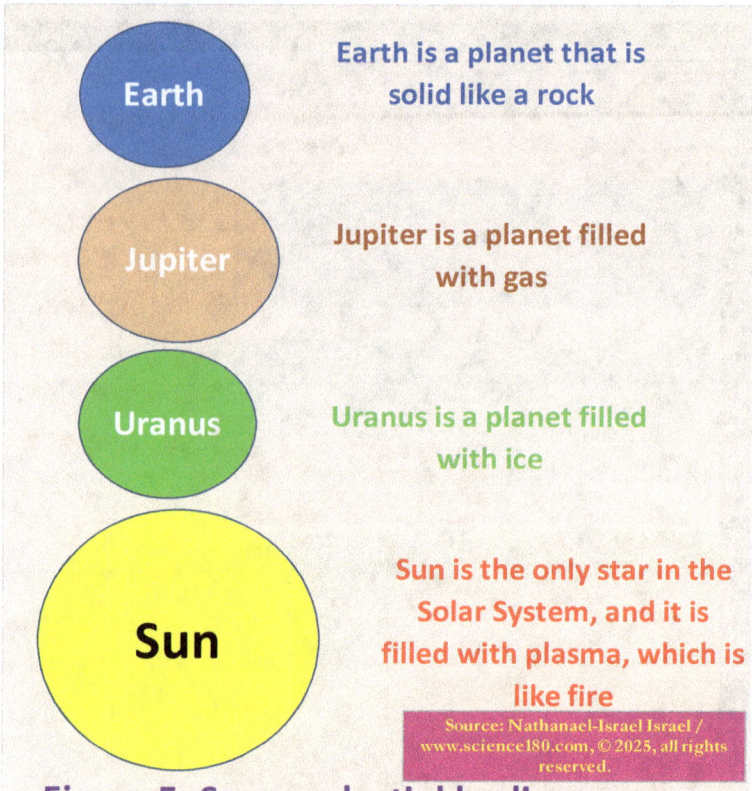

Earth is a planet that is solid like a rock

Jupiter is a planet filled with gas

Uranus is a planet filled with ice

Sun is the only star in the Solar System, and it is filled with plasma, which is like fire

Figure 5: Some celestial bodies are gas, others are ice, solid, or filled with fire.

The Sun is not a planet but a star. It is filled with very HOT materials like lava, fire, steam, and something Daddy calls plasma. Lava and plasma are like the hot material that flows out of the Earth from volcanoes. Daddy told me that each planet has some lava inside. Although it shines, the Moon is not a star. It receives light from the Sun and sends it back to Earth. In fact, just as a light hitting a mirror is pushed or thrown back into another direction, so also when the light coming from the Sun hits the surface of the Moon, it is sent back to the Earth. You can do this experiment by shining a flashlight onto a mirror. You will see how the light is sent in another direction. To say that the light is sent into another direction, scientists use the term "reflected." In other words, when the light coming from the Sun hits the surface of the Moon, it is reflected to the Earth.

Unlike us, who stand up straight as we walk, all planets stand straight as they move around the Sun. Some are almost straight, and others are bent or leaning. Some are bent more than others. Some are even upside down, as if they were walking on their heads rather than their feet. Some are even rolling like a soccer ball rolling on the ground. Jupiter is almost straight.

Nathanael-Israel Israel: Known as the World's Most Accurate Universe-Origin Mathematician

Figure 6: Axial tilt of some planets in the Solar System

The Earth is a little bit bent. Uranus moves like it is rolling like a somersault, whereas Venus is upside down as if it were moving on its head, which is weird. Scientists use the big work called "axis tilt" to determine how bent or tilted the planets are. Figure 6 shows the axis tilt of the planets in the Solar System.

HOW BABY UNIVERSE WAS BORN

I asked my dad why he was telling us all of this stuff that is making me think more than the small question we asked in the beginning. My dad told me to cool down, and he will very soon explain to us (meaning my sister, my brother, and me) why the planets are behaving like that, for all of their attitudes or the way they are acting is linked to how they were made.

By the time Daddy reached this part of the story, we were very excited and couldn't wait until he answered the HUGE questions running through our minds:

- Why is the Sun very big and very hot?
- Why are some planets slow and others fast?
- Why are some planets made with ice, some with gas, and some solid like the Earth?
- How come the planets are not the same?
- Why are the planets located at different distances?
- How did the planets get there?
- Where did they come from?
- Why does the Sun contain some hot materials like lava, and why do some planets also have the same on the inside?
- Are the planets, which are cold, made of ice cream that we can eat?

As I asked those questions to Daddy, he smiled at me and said: "*You are a very smart girl.*" He added that those questions are exactly part of what needs to be explained to address how the Solar System was formed to kids my age. By this time, I felt very happy because I knew Daddy would explain to us how everything in the universe was created. I hope you are following what I am saying, because very soon I will explain everything to you.

My daddy has written many books about how the universe was made, but they are too difficult for a little child like me to understand. That is why he decided to break the story down for us in a simple way that we could understand. Now, I am going to share with you what we learned. Daddy did not tell me everything that happened during the formation of the universe, but what a child like me can understand without having to learn difficult things that people who went to school for many, many years have learned. If you are ready, let's go!

Nathanael-Israel Israel: Known as the World's Most Accurate Universe-Origin Mathematician

4. BEGINNING OF BABY SOLAR SYSTEM

Everything in the universe has a beginning. Just as a baby is born, grows, lies still, crawls, walks, runs, and jumps before starting school and learning different things, so also the universe was born at one time and had to go through different changes so that everything in it could be formed and well-shaped as it grew. Just as a baby grows in height, weight, and strength, so also did things when they were being formed in the universe; they changed and took different forms.

Daddy told me that when he studied the universe, he realized that certain things are not the same or different by chance. For example, it is not by chance that the Sun is so big and contains hot things like lava, while inside the Earth, there is also lava that can produce volcanoes. All of this is because, at the beginning of the Solar System, there was something that looked like a Baby Solar System, which was very hot. Baby Solar System was a gigantic baby that was born and grew into the Solar System. By Baby Solar System, I am not talking about a small baby that you can hold in your hands, but a big thing that contained fires and, at one point, looked like a liquid or gas, let's say something Daddy called a fluid or plasma. A fluid is a liquid, such as water, or a gas, such as air. Unlike babies of human beings, who are small, Baby Solar System was very huge, and deep, and at one point, it was rich in something looking like water. Baby Solar System was not a human being, but something that went through many changes before becoming the Solar System. As Daddy continued the story, he said that, as it was being born, Baby Solar System was pushed by something and started moving very fast.

5. MOTHER SOLAR SYSTEM BIRTHED TWO BABIES

After a certain time, Baby Solar System grew up and became a mother who was pregnant and about to give birth to two children. Suddenly, Mother Solar System split into two parts: one became Baby Sun, and the other became the baby that would become every celestial body orbiting the Sun. In other words, Mother Solar System birthed two babies: Baby Sun and the baby of every celestial body moving around the Sun (see Figure 7). Baby Sun is the thing that grew, went through changes, and became the Sun. In the same way, the other baby that Mother Solar System birthed went through many changes and birthed all the bodies turning around the Sun.

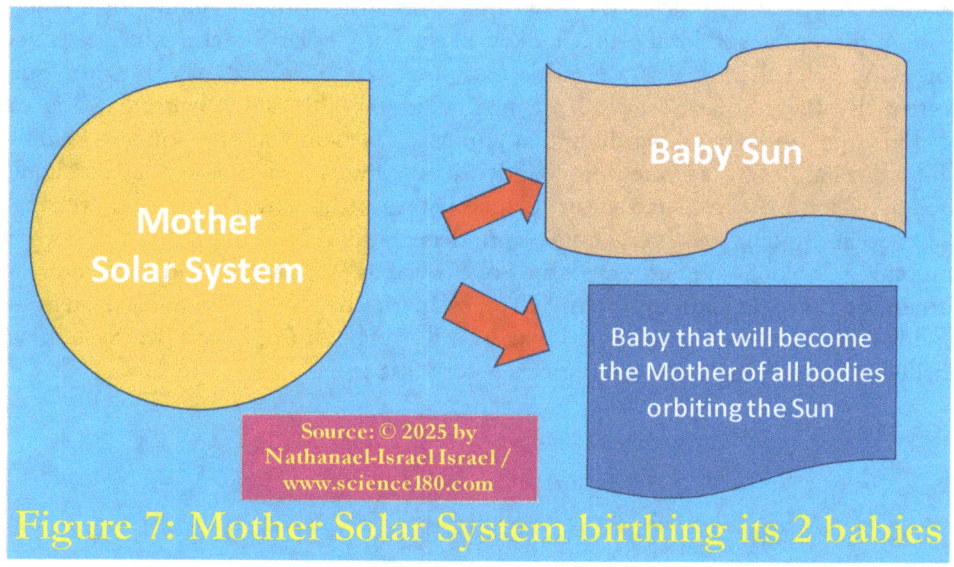

Figure 7: Mother Solar System birthing its 2 babies

All I am saying is just like in a family, a mommy gives birth to a baby who grows until it becomes an adult that gets married and gives birth to other children, and those children will also grow until they get married and give birth to other children. In the end, the family keeps growing, and we can end up having a grandpa who has children, grandchildren, great-grandchildren, and great-great-grandchildren. In the same manner, the babies of the bodies orbiting the Sun grew and birthed other babies. What I am saying about people in the family also applies to the celestial bodies in the Solar System. This is because Mother Solar System gave birth to two babies, which also, in their turn, grew up and birthed other babies. In fact, because Baby Sun was very huge, by the time it went through all of its changes, it had become a very huge Sun. I will explain those changes later. Before I get there, let me tell you something very important about the other Baby of the Mother Solar System.

SECTION 2: HOW THE GALAXIES, THE PLANETS, THE MOON, AND THE SUN WERE FORMED

The other baby was not as big as Baby Sun. That Baby and Baby Sun were both in the belly of Mother Solar System.

As Daddy reached this level of the story, I, Josephine, asked, "Daddy, how does a planet give birth to a baby when it does not have a belly?" Daddy answered by saying that when we talk about babies and mothers here, we are not talking about how a pregnant mother carries a baby in her womb for about 9 months before giving birth. But we are talking about how something like a liquid can break up or change forms to become another thing, which can also go through other changes to become something else. To demonstrate what he was saying, Daddy took us outside our house for an experiment on how water can break up into droplets. Keep reading to learn about this cool experiment.

6. A VERY IMPORTANT EXPERIMENT OF HOW WATER THROWN INTO THE AIR BREAKS INTO MANY MOVING WATER DROPS

How celestial bodies were born is different than how human beings are born. In fact, as baby celestial bodies were moving in space, one of the things that happened to them was that they turned around, and some pieces of the liquids they contained came off. To explain to us what happened during the formation of the universe, Daddy took a glass of water, took us outside, and threw the water into the air so we could observe or see what happens to it. We saw how the water broke into pieces, which moved in the direction Daddy threw it.

To be sure we understood the experiment, Daddy gave each of us a cup of water and a bucket of water, and let us put water in the cups and throw it into the air. We realized that when we threw the water into the air, it arched like a rainbow and broke into pieces that continued to move. Each piece had its own size and moved in its own direction according to the direction into which we threw the liquid. When we threw the water faster, the drops moved faster. We also noticed that some drops are bigger than others. Some drops went very far, and others feel close to us.

When Daddy took us outside for the water experiment, the water in the cup was like the mother water, which, when thrown into the air, gave rise to different water drops, which Daddy called Baby waters. It is almost like taking bubbles and blowing them from the wand into the air. Many Baby bubbles are born and move in the direction they were blown until some fall and others pop. In the same way, when the Mother Solar System was moving, it broke down into Baby Sun and into the Baby that will become all of the bodies turning around the Sun.

Just as some water drops were bigger than others, so also was the size of the babies of Mother Solar System different. In fact, the size of Baby Sun was more than 1000 times bigger than the size of the other Baby of the Mother Solar System. To make things easy, let's call Baby Sun, Baby 1, and let's call the Baby of all the bodies orbiting the sun Baby 2. In other words, Mother Solar System gave birth to two babies: Baby 1, which became the Sun, and Baby 2, which became the mother of all the bodies turning around the Sun (Figure 8).

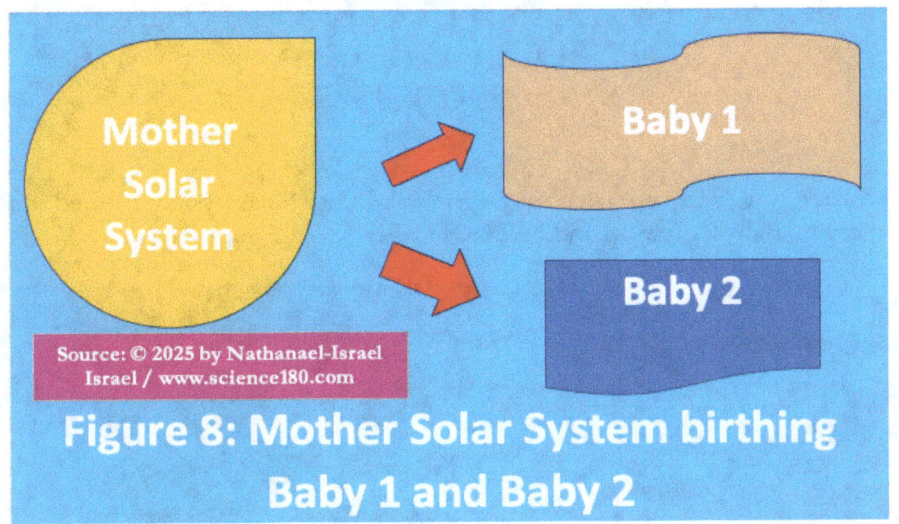

Source: © 2025 by Nathanael-Israel Israel / www.science180.com

Figure 8: Mother Solar System birthing Baby 1 and Baby 2

7. BABY 2 GREW UP AND BECAME MOTHER 2, THE MOTHER OF ALL THE CELESTIAL BODIES ORBITING THE SUN

In the same way, Baby 2 started moving after Mother Solar System birthed it. As it was moving, Baby 2 grew up and became Mother 2 (see Figure 9).

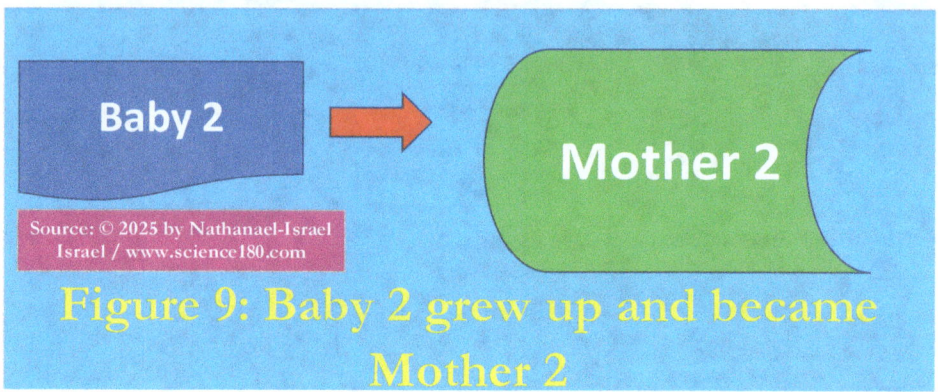

Source: © 2025 by Nathanael-Israel Israel / www.science180.com

Figure 9: Baby 2 grew up and became Mother 2

After Mother 2 traveled for a certain distance at a certain speed, it started giving birth to other babies. One of the first babies Mother 2 birthed was Baby Mercury (see Figure 10). This also explains why, in the end, Mercury is the planet closest to the Sun. After birthing Baby Mercury, Mother 2 continued moving away from the Sun.

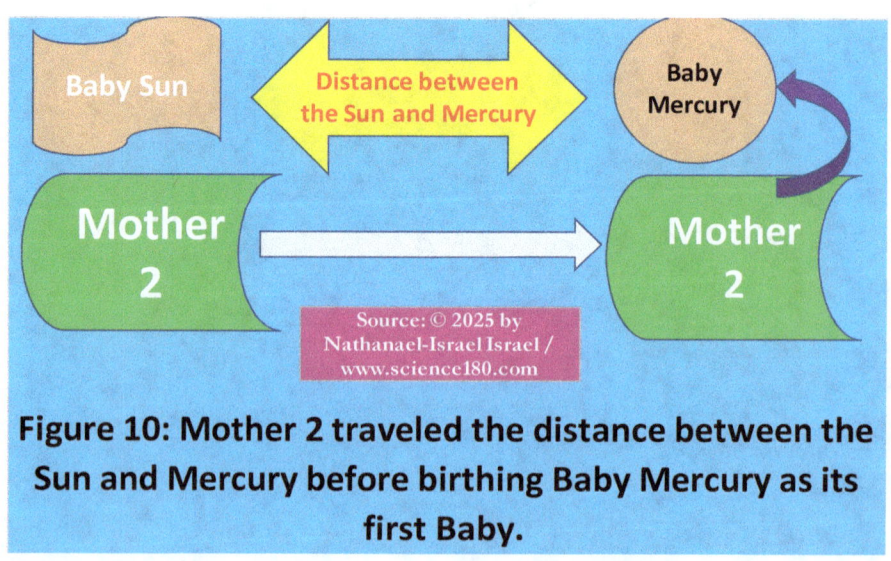

Source: © 2025 by Nathanael-Israel Israel / www.science180.com

Figure 10: Mother 2 traveled the distance between the Sun and Mercury before birthing Baby Mercury as its first Baby.

SECTION 2: HOW THE GALAXIES, THE PLANETS, THE MOON, AND THE SUN WERE FORMED

When Daddy said this, I asked him, "Why did Baby 2, which became Mother 2, have to move away from Baby Sun?"

Daddy replied: "Do you remember the experiment we did for the water we threw into the air, and water drops were formed and moved?" I replied, "Yes."

Daddy then told me that because Baby 2 was pushed away by Mother Solar System at a very high speed, it kept moving, and all of the babies it birthed after becoming Mother 2 also kept moving. Daddy continued the story by saying that Mother 2 kept moving away from Baby Sun, giving birth to different babies until it reached a point when it gave birth to another baby called Baby 3, which grew up to become the Mother of the Earth and the Moon. To make it simple, let's call Mother 3 the Mother of the Earth and the Moon. In other words, Mother 2 gave birth to Mother 3. As a reminder, Mother 2 was the mother of ALL celestial bodies turning around the Sun. The distance that Mother 2 traveled away from Baby Sun before giving birth to Mother 3 is about the distance separating the Earth from the Sun. Figure 11 shows that.

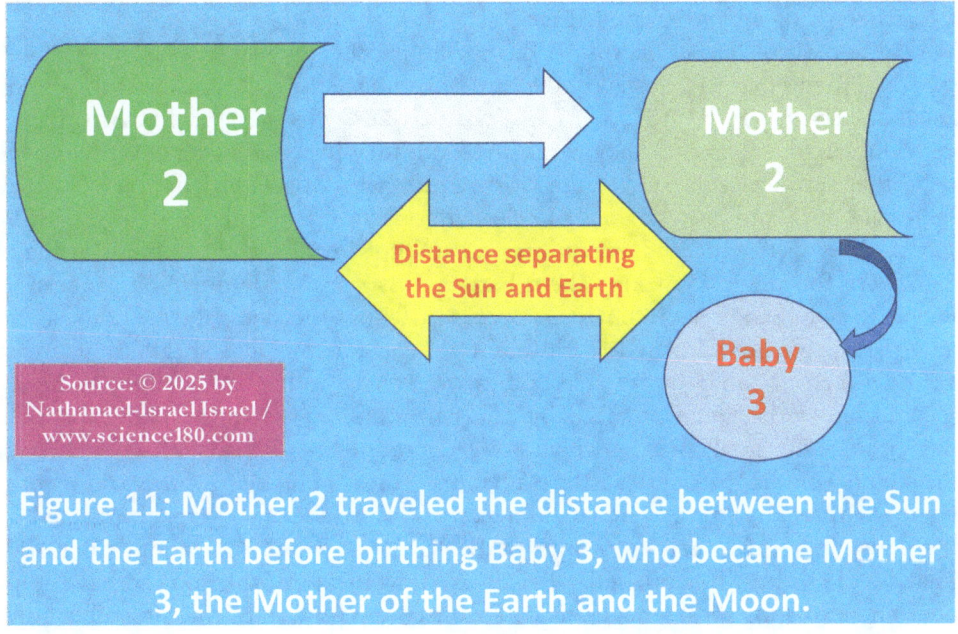

Figure 11: Mother 2 traveled the distance between the Sun and the Earth before birthing Baby 3, who became Mother 3, the Mother of the Earth and the Moon.

Daddy taught us that, using the speed at which Mother 2 moved and the distance between the Sun and the Earth, he calculated how long it took for Mother 3 (Mother of the Earth and the Moon) to form. We will get back to that later. For now, let's see how Mother 3 gave birth to its children.

Science180: All the Universe-Origin and Life-Origin Solutions You Love

8. MOTHER OF THE EARTH AND THE MOON BIRTHED BABY EARTH AND BABY MOON

The Mother of the Earth and the Moon, which Daddy called Mother 3, quickly gave birth to Baby Earth and Baby Moon. Baby Moon was much smaller than Baby Earth and was pushed away from Baby Earth. Baby Moon traveled for a certain distance before becoming the Adult Moon. Based on the great research that Daddy has done, the distance that Baby Moon traveled before forming the Adult Moon is about the distance separating the Earth and the Moon. That is why the Moon is far from Earth. After Baby Moon moved away from Baby Earth, Baby Earth went through some changes and became the Adult Earth that we call Earth today. Likewise, Baby Moon underwent some changes and became the Adult Moon we call the Moon today.

At this time of the story, my sister Joelle-Major raised her hand and asked Daddy: "What changes did Baby Earth, Baby Moon, and Baby Sun go through before becoming what they are today?"

Daddy replied that, unlike the babies of human beings who are born small and grow big, the Babies of celestial bodies are born huge and grow small. At this point, we all smiled and said: "WHAT?! How can that be?" We thought that ALL babies are born very small, and as they eat more food, they become bigger. Daddy replied to us that how celestial bodies are born and grow is different than how human Babies are born and grow. Babies of celestial bodies were born big, but as they were moving, they were squeezed or compressed and got smaller as small things inside of them were forming and making them harder. It is not exactly how you squeeze an orange to get orange juice, but it is because of how the liquid in the Babies was being organized inside to form different things, looking like spaghetti wrapped around a fork, as the Baby celestial bodies were trying to become adults (see Figure 12). In fact, as the liquid of the baby celestial bodies was trying to come together, they broke into small pieces, which moved around in circles like whirlwinds or tornadoes, spinning around at the same time they were being pressed to come together. In the end, they were small pieces of water here and there, compacted on their own, but put together as one body. "That is weird!" I responded. Daddy replied: "That's how it happened." This is like how inside a fruit, you can have some seeds and also some fleshy materials you can eat, and how all of them together form the fruit.

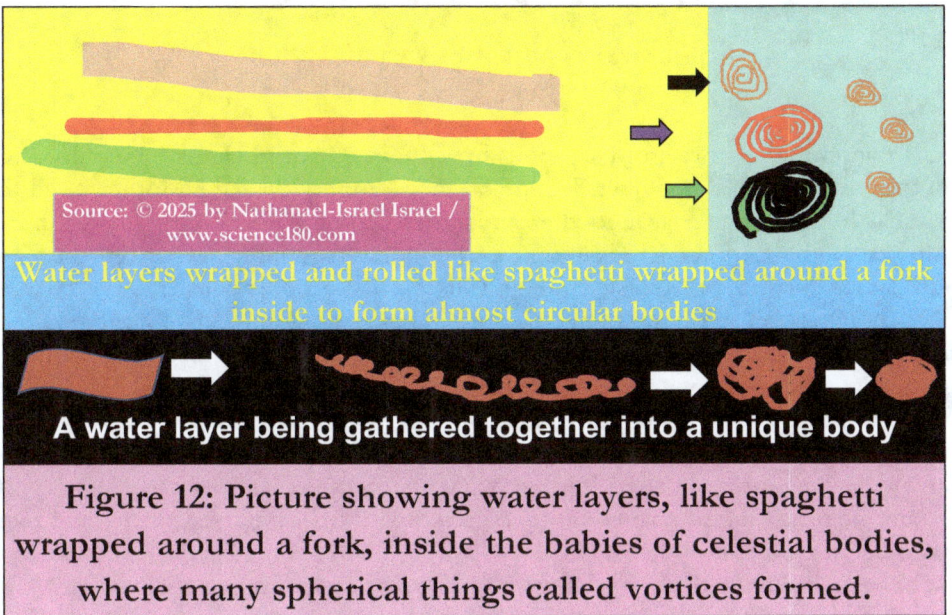

Water layers wrapped and rolled like spaghetti wrapped around a fork inside to form almost circular bodies

A water layer being gathered together into a unique body

Figure 12: Picture showing water layers, like spaghetti wrapped around a fork, inside the babies of celestial bodies, where many spherical things called vortices formed.

9. HOW MARS, JUPITER, SATURN, URANUS, NEPTUNE, AND PLUTO WERE FORMED

Let's continue the story and talk about how Jupiter, Saturn, Uranus, Neptune, and Pluto were formed. Remember, I told you that Mother 2 was the Mother of all the celestial bodies turning around the Sun. After Mother 2 gave birth to Mother 3 (Mother of the Earth and the Moon), it continued its journey away from the Baby Sun, giving birth to many asteroids until it reached a point when it gave birth to Baby Mars. After birthing Baby Mars, Mother 3 continued its journey until it reached a point when it birthed Baby Jupiter. The journey continued until Mother 3 reached another point where it gave birth to Baby Uranus. After Baby Uranus, Baby Neptune was born, and later Baby Pluto.

All of those babies grew up and became planets. For instance, Baby Jupiter grew up and became Jupiter. Baby Saturn grew up and became Saturn. Baby Uranus grew up and became Uranus. Baby Neptune grew up and became Neptune, while Baby Pluto became Pluto.

Before I continue the story, let me remind you of what we have said so far. After Baby Solar System was born, it grew up and became Mother Solar System. Then, Mother Solar System birthed two children: Baby 1, which grew up to become the Sun, and Baby 2, which grew up to become Mother 2, meaning the Mother of all the celestial bodies orbiting the Sun. We also learned that, as it was moving away from Baby Sun, Mother 2 birthed many children until it reached a certain point when it birthed Baby 3, which became Mother 3, the Mother of the Earth and the Moon. I hope you remember all that.

As Mother 2 was moving away from the Sun and giving birth to many children, it got older and smaller until it could no longer give birth to children anymore. After Mother 2 birthed all of its children, it died. Each of those children grew up, went through changes, and became adults. Some of those Adult children of Mother 2 gave birth to their own children. That is how some children of Mother 2 ended up giving birth to a planet orbited by (meaning moving around, which turned) some satellites. As a reminder, a satellite is a celestial body that orbits a planet, just like how the Moon orbits the Earth.

l body that turns around a planet, just like how the Moon turns around the Earth.

10. WHY DOES THE EARTH TURN AROUND THE SUN WHILE THE MOON TURNS AROUND THE EARTH?

As Daddy was saying this, I raised my hand and asked, "Why does the Earth turn around the Sun while the Moon turns around the Earth?

"*Great question*!" said Daddy. This is about how celestial bodies were formed in the universe. To explain this, Daddy talked to us about what is called a planetary system: a family of celestial bodies, including a planet and its satellites. In fact, each planetary system was a baby at one point. In other words, it was a baby planetary system that grew up to become a mother that birthed a Baby planet and Baby satellites.

When each Baby planet was born, it was positioned almost in the middle of what would become the planetary system and started turning around the Baby star. The way the Baby satellites were born caused them to start orbiting, meaning turning around, the planet in their planetary system. That is why, until today, all planets turn around a star and all satellites turn around a planet. Because the Sun is the star in the Solar System, all planets in the Solar System orbit the Sun. Likewise, all satellites turn around their planet. Hence, the Moon orbits the Earth. Jupiter, for instance, has more than 80 satellites, all of which orbit it. Do you understand? I hope so!

11. WHY DO SOME PLANETS WALK ON THEIR HEAD, WHILE OTHERS ROLL LIKE A BALL, AND SOME ARE MORE BENT THAN OTHERS

When Mother 2 was moving away from the Sun, it took a certain time before it birthed each of its babies. That is why the planets in the Solar System are not together but are separated by a huge distance. As those babies were being born, those who were born first moved faster than those who were born last. Some of those babies were holding hands as they were being born, and when their hands were separated, they were pushed into different directions, just as two people holding hands can be sent into two different directions when the hands are released. In the case of the celestial bodies when the hands were removed, some fell backwards. That is why some celestial bodies are walking on their heads, and others are tilted or oriented differently. Some of them, like Jupiter, are not flipped much because they are very heavy. When these babies were being born, they spun around or rotated; hence, in the end, all celestial bodies turn around all of the time. Some rotate very fast, while others are slow. For example, the Earth spins around completely once every 24 hours. Some planets take many months to complete a single rotation. During its spinning around, the Earth faces the Sun in the day, while in the night it does not face the Sun.

12. WHY DO PLANETS HAVE DIFFERENT SIZES AND COLORS?

Recalling what I have learned at school, I asked Daddy the following questions:

- Why are some planets gas, others solid, and some icy?
- Why do planets have different colors?

Planets have different colors because they are made of different chemicals. Mars usually looks red. Jupiter looks like an orange-and-white cloud. From space, the Earth looks blue.

Daddy told us that, just as human beings have different sizes, so also celestial bodies were born with different sizes. Some celestial bodies are very small, and others are very BIG. Jupiter and Saturn are the biggest planets orbiting the Sun. This is because their babies were very big.

All children of celestial bodies were smaller than their mothers. That is also why the water drops we obtained by throwing a ball of water in the air were smaller than the water we initially threw. Human beings give birth to children who can grow and become as big as, or even bigger than, their mother. But in the case of celestial bodies, the babies are always smaller than their mother. This is because a mother of celestial bodies had to break into many pieces before birthing its children.

13. WHY ARE SOME PLANETS SOLID, WHILE OTHERS ARE GAS OR ICY?

When Mother 2 was moving away from Baby Sun, its temperature decreased, which means its temperature went down, as it was moving away from Baby Sun.

Just as water put into a freezer can become ice, so also the children of Mother 2 that were born far away from the Sun were rich in ice. Some children born to Mother 2 are rich in gas, while others are solid, like the Earth. For example, Jupiter and Saturn are gas giants. Uranus and Neptune are filled with ice, while Mercury, Venus, Earth, Mars, and Pluto are very hard. It all boils down to how the Babies of Mother 2 were pressed down, compressed, or squeezed as they were being shaped. For instance, a slice of bread is very soft. But if a slice of bread is squeezed or compressed, it can become harder or more solid. The same thing happened when celestial bodies were being formed. Some babies of Mother 2 were squeezed to become solid like the Earth, while others were not that squeezed and became gas planets like Jupiter. In other words, some of the babies of Mother 2 were highly squeezed to birth a solid planet like the Earth, while others were less squeezed and birthed gigantic planets like Jupiter.

The temperature of the environment also played a role in what the babies' celestial bodies became. To illustrate or better explain what he was saying in a language that we children could understand, Daddy reminded us that the same water we drink when we eat can also become solid like ice, and also be gas like the steam when we cook food. In other words, water can be in the form of ice, liquid, or gas. If water is put in a very cold environment like a freezer, it freezes and becomes ice. If water is left at room temperature, it stays liquid. When water is boiled on the stove, it can evaporate, meaning it goes into the atmosphere like a gas that we can see in the form of water steam. Others were born in very cold environments and became rich in ice like Neptune.

Nathanael-Israel Israel: Known as the World's Most Accurate Universe-Origin Mathematician

14. WHY DO PLANETS HAVE DIFFERENT SPEEDS, AND WHY DO THOSE CLOSE TO THE SUN ORBIT THE SUN FASTER THAN THOSE FAR FROM THE SUN?

Before I say something very important about this question, let me remind you that Mother 2 was the mother of all the celestial bodies orbiting around the Sun. As Mother 2 was moving away from Baby Sun, it was getting tired, and the babies it was giving birth to as it got older and older were moving more slowly. Part of this is because when Mother 2 was pregnant with its babies, they were stacked one on top of the other like pancakes and moved along a current like how a river flows. In fact, in a river, waters are stacked on top of each other, and they form water layers, like pancakes laid one on top of the other in a moving stream or in a flowing river. You may not know it, but it is very true that all the water in a river doesn't move at the same speed. In fact, in a river, the waters on top moved faster than those beneath them. In the end, the waters at the bottom move more slowly than any above them (see Figure 13).

1	Mercury	Baby of the planets showed as water layers stacked on top of each other. Because the layers at the bottom carried more weight, they moved more slowly and gave rise to planets that orbit the Sun more slowly than those born from the top layers.
2	Venus	
3	Earth	
4	Mars	
5	Jupiter	
6	Saturn	
7	Uranus	
8	Neptune	

Source: © 2025 by Nathanael-Israel Israel / www.science180.com

The Baby planets carrying different water loads on their head. The Baby planets on top of the water layers of Mother 2 carry less load than those at the bottom. Baby Mercury carries no load, Baby Venus carries 1 load, Baby Earth carries 2 loads, Baby Mars carries 3 loads, Baby Jupiter carries 4 loads, etc. Hence, in the end, the planets close to the Sun move faster than those far from the Sun.

Figure 13: Why planets have different speeds, and some move faster than others

Likewise, the first babies Mother 2 birthed were like the first pancakes or the first layers of water on top of the moving stream. Those babies moved faster. The pancakes on top pressed down on the pancakes below, and because the bottom ones were pressed, they carried the weight of those on top. In other words, the pancakes on the bottom moved slowly because they carried too much weight. That is why the first Baby Mother 2 gave birth moved faster than the last Baby. For instance, Mercury is the closest planet to the Sun, and it moves faster than any other planet in the Solar System. Venus, which was born after Mercury, moves more slowly than Mercury. Then the Earth, which was born after Venus, moves more slowly than Venus does. Mars was born after the Earth, and as you can imagine, Mars moves more slowly than Earth. Likewise, because Jupiter was born after Mars, Jupiter moves more slowly than Mars. Here, when I talk about movement, I mean how the planets orbit the Sun. The speed of the planets around the Sun is not about how big they are, but about how close or far they are from the Sun (Figure 13).

To wrap it up, the celestial bodies that are close to the Sun moved fast because they were born from the pancakes or water layers that were closer to the top of the water of Mother 2, but the celestial bodies that are farther away from the Sun move slower because they were born from the pancakes or water layers that were at the bottom of Mother 2. That is why the speed of the celestial bodies decreases from the closest body to the Sun all the way to the last body turning around the Sun.

15. WHY DO PLANETS ROTATE?

At this point. Daddy asked us if we wanted to know "*Why do planets turn around?*"

Joshua-Enoch replied "*No*" because he never knew that planets turn around, and that Daddy was teaching us was too much information for his head to understand, and he was ready for a break. But Joelle-Major and I told Daddy that we wanted to know why some planets turn around themselves. Then, Daddy asked my siblings if they knew what rotation meant.

Before he introduces a new topic to us, Daddy likes to know what we know about it, so he knows where to start his teaching. That is why Daddy likes to ask us questions. Only I knew about the rotation of planets.

To help explain what turning around or rotation is, Daddy took us to the dining room and pulled all of the chairs out from the circular dining room table. Daddy first made all of us walk around the table. As we were walking, he told us our movement at that time was like a celestial body orbiting the Sun, meaning a celestial body moving around the Sun. In other words, when a planet moves around the Sun, we say that it orbits the Sun.

Then, Daddy told us that celestial bodies do not move around the Sun like we walk around the dining room table. But the celestial bodies spin around themselves as they move around the Sun. To help us better understand how rotation works as planets orbit the Sun, Daddy told us to spin around while we went around the table. We started spinning in circles and orbiting the table. It was a cool experiment, and we started laughing. We quickly became very dizzy and had to stop, or we would have fallen. Therefore, Daddy told us to stop moving. He then told us that spinning around in circles is called turning around or rotation, while turning around the table is called revolution or orbital movement.

Then, Joelle-Major asked: "*How come we spun around for just a few minutes and were very dizzy and about to fall down, yet the planets have been spinning around themselves as they orbit the Sun for thousands of years, but they never get dizzy and fall down as we were about to fall down?*"

As she asked that question, I followed up and asked Daddy, "*Why do we get dizzy when we spin around?*" Daddy laughed and said that our questions are great and that we are very smart kids! He was also happy because he knew we were paying attention to what he was saying and that we were very curious.

Daddy then told us that when we turn around or spin for a long time, our bodies are shaken too much. Because that movement shakes our body and the brain in our head, it ends up affecting our whole body. The brain is something like a liquid containing some well-organized stuff, and when they are shaken, this stuff can be scrambled just as the yellow and white of an egg get messed up when they are scrambled.

Going back to Joelle-Major's question about why celestial bodies rotate like a tornado, Daddy said it all boils down to how they formed. He reminded us that the

SECTION 2: HOW THE GALAXIES, THE PLANETS, THE MOON, AND THE SUN WERE FORMED

babies of the celestial bodies once moved like a stack of pancakes or the layers of water in a moving river. When water moves in a river, it is stacked on top of other water. As they move, the top layers flow faster than the bottom layers. As those layers of water in the baby planets were moving, they started collecting themselves in circles and rolling around just like how we can roll spaghetti around a fork. For example, a long spaghetti noodle can be rolled around a fork until it forms a solid thing around the fork. It is like a piece of yarn that you roll around into a ball. In the same manner, the water or liquid of the baby of the planet rolled around itself until it formed the planet, and the rolling continued in the form of the orbital movement and rotation, or the spinning around. That is why planets spin on their axes as they orbit the Sun. Daddy told us that in the books he wrote for adults, he had to better explain this very difficult subject called turbulence, something that is way beyond what children's minds could understand. Turbulence involves things that also explain how clouds are formed in the atmosphere and can produce heavy rain, like hurricanes or tornadoes. Now that we understand how the Sun and the planets were formed, let's move to other things.

Science180: All the Universe-Origin and Life-Origin Solutions You Love

16. HOW WERE THE GALAXIES FORMED?

When Daddy asked us in the beginning of this book to write down our questions, one of the things that my sister Joelle-Major wanted to know was how galaxies were made. Daddy was a little surprised that she even knew what the word "galaxy" means. Although most children our age may have heard about the word "galaxy" and thought of it as a cool thing in the sky, Daddy was not sure if we really knew what a galaxy meant. Therefore, he started answering this big question by telling Joelle-Major: "*I know you heard about galaxies somewhere, but do you really know what a galaxy means?*"

All of us were sitting around the dining table, and Daddy was answering our questions and typing them at the same time. We started with Joshua-Enoch's questions, and Daddy answered them one by one in the order they were written down. Then, he answered the question of Joelle-Major next, and he finished with mine. As he was answering the questions, he told us if any of us did not understand anything, to stop him and ask. Obedient to his instructions, we stopped every time he said anything hard to understand or even when he said a word that was for adults. We asked many questions that he answered for us, but he did not include all of them in this book, for he said that if he were to include them all, it would be a huge book that people would not enjoy. Also, some questions were not about the books' topics. Before answering any question, he usually asked us what we knew about it. This gave us a chance to talk and speak about what was on our minds before Daddy improved it or corrected it.

Therefore, as I felt like Joelle-Major did not know what galaxies mean, although she was the one who asked that question, I quickly came to her rescue by saying that galaxies are made of many stars that are shaped in a special way, as if they were stuck together somehow.

"*That is a pretty good answer,*" Daddy replied. Galaxies are made of many stars organized into groups. In other words, the stars in a galaxy are bound together like people in the same family.

As a reminder, the Solar System is a group of celestial bodies made of the Sun at the "center", and the planets and asteroids orbiting or turning around the Sun. Most stars belong to a specific galaxy. There are billions of galaxies in the universe, and most of them have not been discovered yet. The Sun and all the other celestial bodies in the Solar System belong to a galaxy called the Milky Way Galaxy. As soon as Daddy said the word "Milky Way," I asked him why they call our galaxy the Milky Way, when it is not made of milk.

"*Very smart girl! Josephine!*" Daddy replied to me. Daddy always encouraged us to ask questions, and he was very happy when we asked difficult questions. Hearing Daddy call me smart, I smiled and looked at him. Then, Daddy said that our galaxy is called the Milky Way because it looks like milk or a stream of milk! In addition, Daddy told us that all the stars we see in the sky at night belong to the Milky Way

SECTION 2: HOW THE GALAXIES, THE PLANETS, THE MOON, AND THE SUN WERE FORMED

galaxy. The Milky Way galaxy looks like a spiral. Then, Daddy asked us if we knew what a spiral meant. I answered that a spiral is like spaghetti wrapped around a fork. Joshua-Enoch added that a spiral is like a snake wrapped around a leg. Because Daddy does not like snakes, he asked Joshua-Enoch to try again. This time, Joshua-Enoch said that a spiral is like ternary spinning around.

"*Good try*," Daddy replied.

The universe contains many galaxies. Just as the Solar System was once a baby we called Baby Solar System, so were all the stars in the universe, along with their planets and asteroids. In the same manner, all the galaxies in the universe were once baby galaxies, which had to grow up to birth many babies who grew up and became stars orbited by planets and asteroids. In other words, all the galaxies in the universe were initially babies that grew up to give birth to all the stars, planets, asteroids, and other celestial bodies they contain. If you remember, I told you that the Sun is filled with fire, something like lava that is found in volcanoes, and also something called plasma. Similarly, all Baby galaxies were filled with Baby stars popping up here and there like popcorn or fireworks. That means that when galaxies were forming, the universe could have looked like a gigantic firework show.

In response to this statement, I said that Baby Universe could have been like a popcorn show. Everybody laughed at that illustration. Just as a firework pops up in the sky and births other fires, which also can pop, when the universe was being formed, huge Baby galaxies were popping up, giving birth to many stars everywhere. This was more than the July 4th fireworks that the Americans put up to celebrate their independence from the UK. Anyway, during the formation of the universe, no human being was formed yet; otherwise, very hot fireworks could have hurt, burned, or even killed them. At that time, no dinosaur, giraffe, or rhinoceros had been made yet. It was after the Earth was formed that animals, plants, and human beings were formed. I will get back to that later in this book.

Very soon, I will explain to you how all living things were made. Before that, let's do some STEAM to see how fast the universe was formed and who did that. I hope you understand that by the steam, I didn't mean the hot steam we use to cook food in our kitchen, but STEAM, which means Science, Technology, Engineering, Art, and Math. Even if you are not very good at science and math, we will help you understand it 100%.

Are you ready? Are you happy? Daddy asked.

With a loud voice, we all replied, "Yes." And he said, "Excellent." Let's go!

Science180: All the Universe-Origin and Life-Origin Solutions You Love

Nathanael-Israel Israel: Known as the World's Most Accurate Universe-
Origin Mathematician

SECTION 3: POWERFUL MATH ABOUT HOW LONG IT TOOK TO FORM THE EARTH, THE MOON, AND THE SUN

17. LET'S REMEMBER WHAT WE LEARNED ABOUT THE SOLAR SYSTEM 'S BIRTH.

So far, we have been talking about how Baby Universe was born, but we have not talked yet about how long it took for it to be born. Daddy told us that when most women get pregnant, the baby stays inside the womb for about 9 months before being born.

"Why does a baby have to stay in mommy's belly for 9 months, while it can be taken out and grow?" I asked Daddy. He replied that all this is because it takes time for a baby to grow before it is ready for birth. After a baby is born, he or she will still need to grow before becoming a teenager, and later an adult.

"How long did Baby Universe take before birthing all the celestial bodies then?" I asked Daddy.

Before using science to address that issue, let's summarize what we have learned so far.

Daddy said that he will answer by first focusing on the Solar System, meaning the group or family of celestial bodies formed by the Sun and all the celestial bodies (including planets and asteroids) orbiting it. By the way, "orbiting" is a fancy way to say that a body is turning around another one, just as the planets in the Solar System turn around the Sun, or how the satellites turn around a planet.

Source: © 2025 by Nathanael-Israel Israel / www.science180.com

Figure 14: Baby stars were popping up like popcorn in Baby Universe

We also learned that when Baby Universe was being born, many Baby stars were popping up here and there in space. It was during that process that Baby Solar System was born (see Figure 14). Then, Baby Solar System grew up to become Mother Solar System (see Figure 15) before giving birth to 2 babies, whom we called Baby 1 and Baby 2 (see Figure 16).

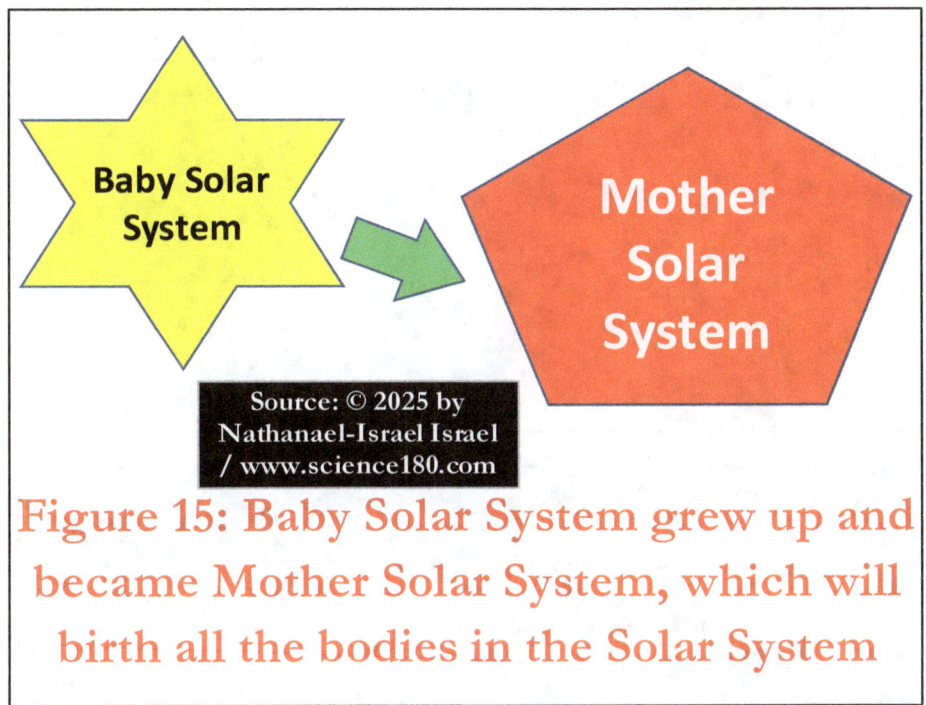

Source: © 2025 by Nathanael-Israel Israel / www.science180.com

Figure 15: Baby Solar System grew up and became Mother Solar System, which will birth all the bodies in the Solar System

Baby 1 was Baby Sun that grew up to become the Sun (see Figure 17). Baby 2 was the baby that grew up to become the Mother of all the celestial bodies orbiting the Sun. In other words, after Baby Sun was born, it was collected into the Adult Sun, while, in turn, Baby 2 grew up to become Mother 2, meaning the Mother of all the celestial bodies orbiting the Sun (see Figure 18).

Figure 16: Mother Solar System birthed Baby 1 and Baby 2

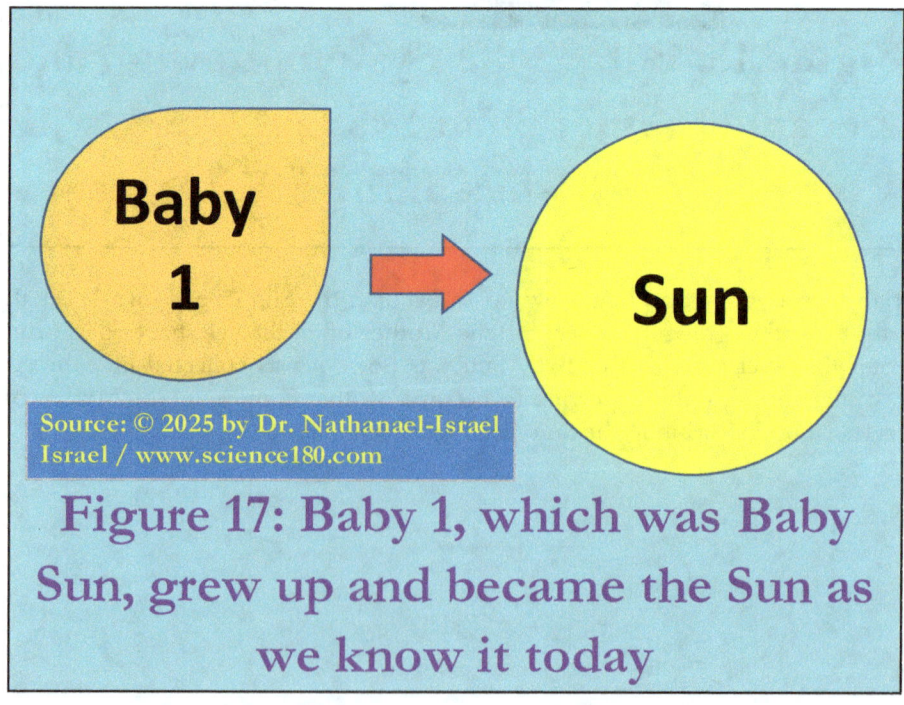

Figure 17: Baby 1, which was Baby Sun, grew up and became the Sun as we know it today

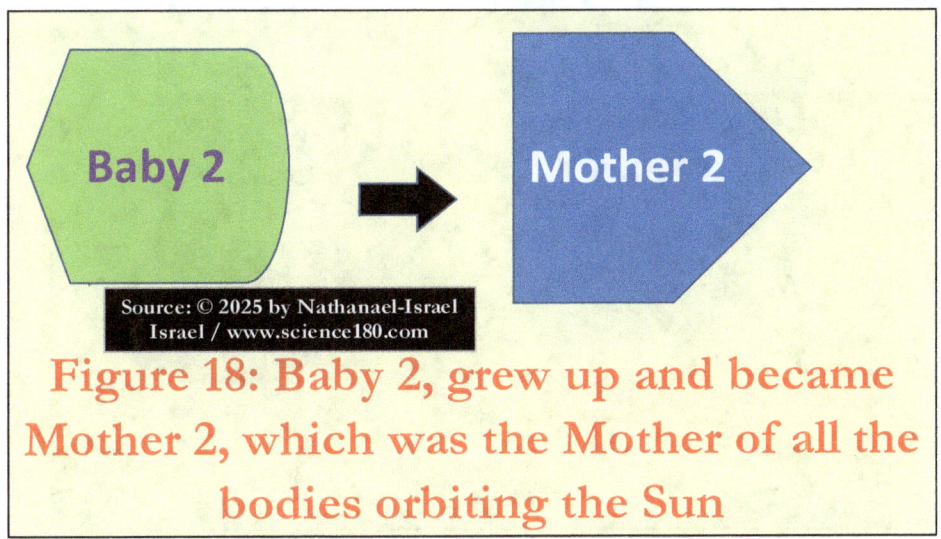

Figure 18: Baby 2, grew up and became Mother 2, which was the Mother of all the bodies orbiting the Sun

As Mother 2 was flowing like a river carrying many babies stacked one on top of the other like pancakes or water layers, each baby was born according to its position in the stack of pancakes. Some babies in the huge stack of Mother 2 became Baby planetary systems and grew up to become Mother planetary systems. By the way, a planetary system is a family of celestial bodies formed by a planet and its satellites. After a Baby planetary system formed and became a Mother planetary system, it then gave birth to a Baby planet and Baby satellites. Mother 2 also gave birth to many Baby asteroids between the Baby planets. The babies that were on top of the pancake stack or water layers were born first, and those that were at the bottom were born last. One of the very first babies Mother 2 gave birth to was Baby Mercury. After Baby Mercury was born, Mother 2 continued its journey away from Baby Sun until it reached a certain point when it birthed Baby 3 (see Figure 19), which quickly grew up and became Mother 3, meaning the Mother of the Earth and the Moon (see Figure 20).

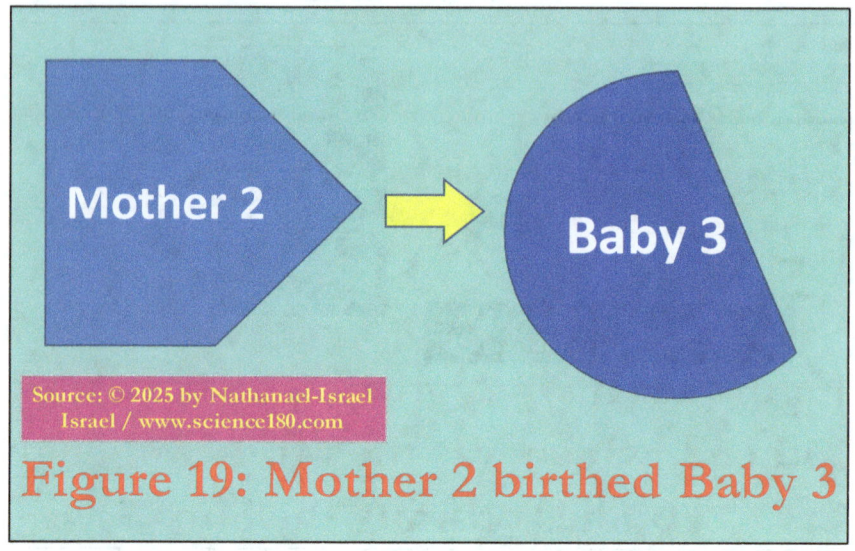

Source: © 2025 by Nathanael-Israel
Israel / www.science180.com

Figure 19: Mother 2 birthed Baby 3

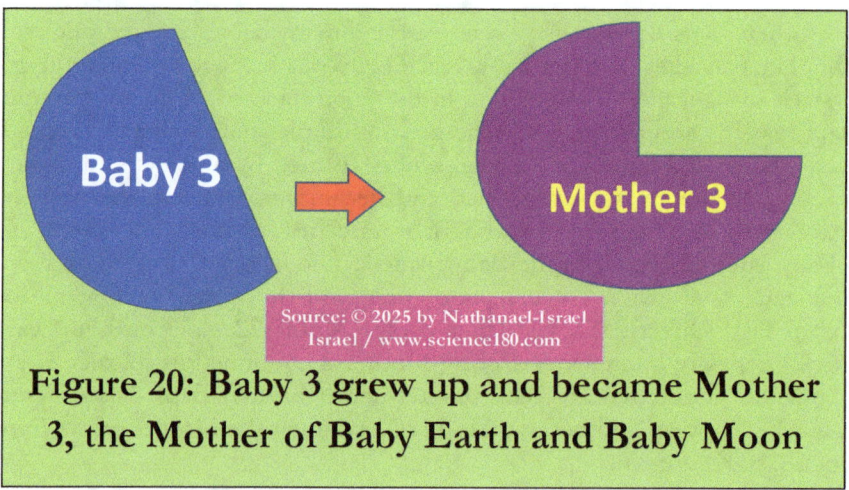

Source: © 2025 by Nathanael-Israel
Israel / www.science180.com

Figure 20: Baby 3 grew up and became Mother
3, the Mother of Baby Earth and Baby Moon

Then, the Mother of the Earth and the Moon fast birthed Baby Earth and Baby Moon (see Figure 21). As you can see in Figures 22 and 23, after some changes to their bodies, Baby Earth became the Earth and Baby Moon became the Moon.

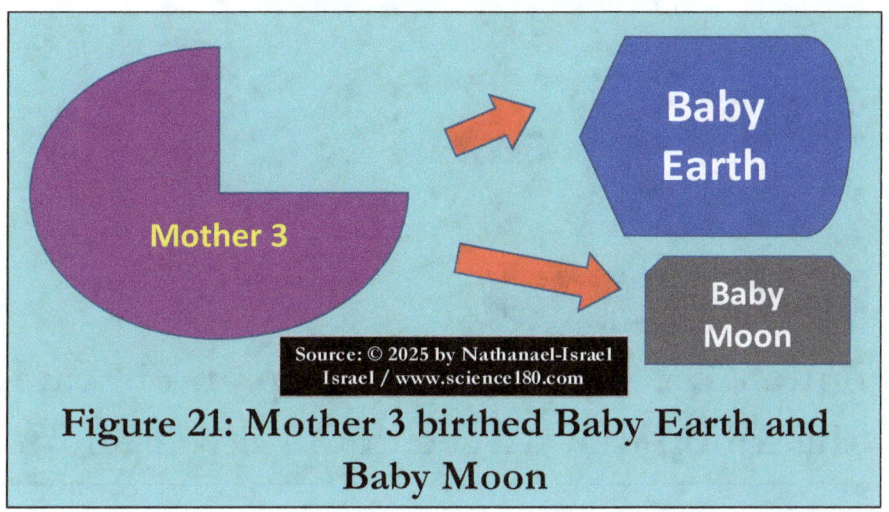

Mother 3
Baby Earth
Baby Moon
Source: © 2025 by Nathanael-Israel Israel / www.science180.com

Figure 21: Mother 3 birthed Baby Earth and Baby Moon

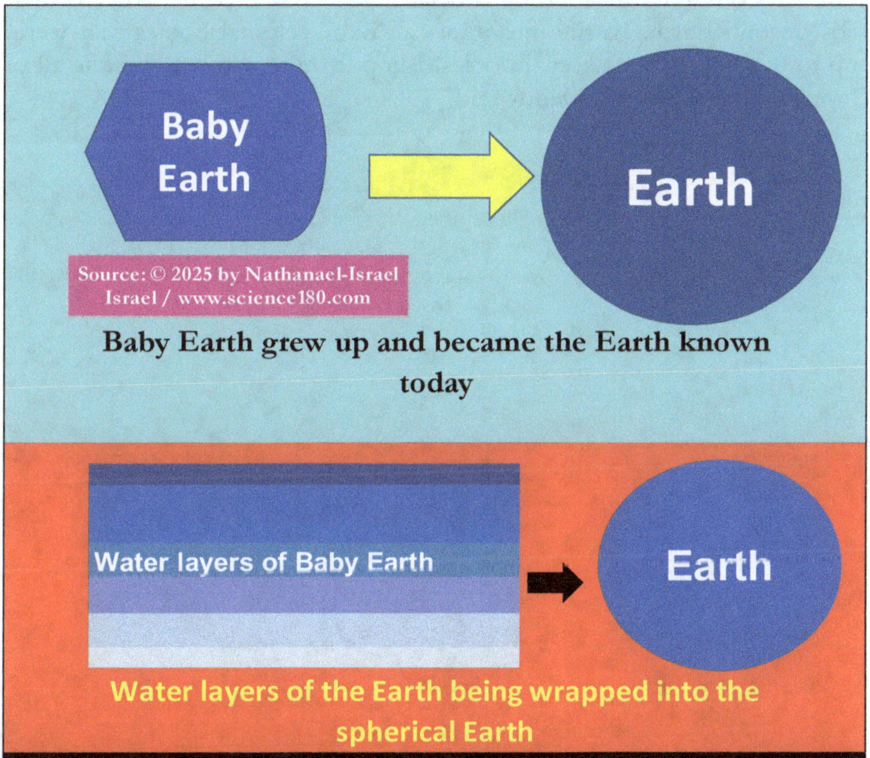

Baby Earth
Earth
Source: © 2025 by Nathanael-Israel Israel / www.science180.com
Baby Earth grew up and became the Earth known today
Water layers of Baby Earth
Earth
Water layers of the Earth being wrapped into the spherical Earth

Figure 22: How Baby Earth became Adult Earth

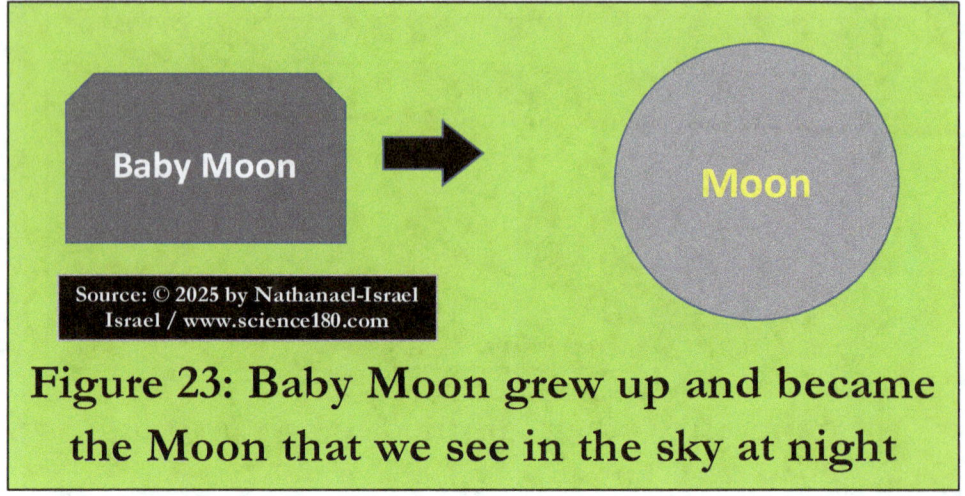

Figure 23: Baby Moon grew up and became the Moon that we see in the sky at night

Mother 2 continued its journey and later gave birth to the other planets and asteroids beyond Earth. In the meantime, all Baby celestial bodies that were born grew up to become Adult celestial bodies. I hope you remember and like all of this. "*Yes,*" we all replied. "*Good*", Daddy said.

18. HOW CAN WE USE STEAM (SCIENCE, TECHNOLOGY, ENGINEERING, ART, AND MATH) TO CALCULATE THE TIME IT TOOK FOR THE EARTH, THE MOON, AND THE SUN TO BE FORMED?

To demonstrate means to prove or to show how that thing happened. For example, demonstrating how long the Earth took to form means showing how long it took before it was formed. We will also demonstrate how long it took for the Moon and the Sun to be formed. Before we do that, I need to explain two important words: distance and speed. People in the US like to measure distance in miles, but people in Africa and Europe prefer to use kilometers. By the way, one mile is about 1.61 kilometers. Let's consider an example.

Because my family's house is close to the elementary school my brother, sister, and I attend (the school we go to), we walk every morning to get to class. From our home to the school is about half a mile. That half a mile is called a distance. In other words, a distance is the length or the amount of space separating two things or people. The distance we walked every morning to go to school is the length separating our home and our school.

To get to the school, we usually walk at a certain pace that works for all of us. Because we have shorter legs than Daddy, we cannot walk as fast as he does. Therefore, he slows down as much as possible so we can all catch up with him without having to run. What a great Dad I have!

We usually take about 10 minutes to walk the half mile separating our home and our school. Because we can walk half a mile in 10 minutes, it means that we can walk a whole mile in about 20 minutes. The math I did to know that we can walk one mile in 20 minutes is called "multiplication." I multiplied 10 minutes by 2 to get 20 minutes, and I also multiplied half a mile by 2 to get one mile. I hope you understand; if not, let me put it in a real scientific way: $10 \times 2 = 20$ and $\frac{1}{2} \times 2 = 1$. By the way, $\frac{1}{2}$ is what scientists write as 0.5.

To speak like a mathematician (meaning someone who is good at math), we will say that we can walk about 1 mile in 20 minutes. Because there are 60 minutes in one hour, if we walk 1 mile every 20 minutes, we can then walk about 3 miles in 1 whole hour, provided we don't stop and take a break. To say that we can walk about 3 miles per hour is to say that our speed is 3 miles per hour, written as 3 miles/hour. In other words, speed is a way to say how much distance you can cover within a certain amount of time. Some people can walk faster; others can run faster. Olympians, athletes who compete at the Olympics, are usually the fastest and set records.

To calculate how long it takes to walk, run, or fly over a certain distance at a certain speed, the math that is usually done is to divide the distance by the speed.

For example, let's calculate how long it will take to walk 6 miles at a speed of 3 miles per hour. Here, the distance is 6 miles and the speed is 3 miles per hour. Dividing the distance by the speed gives the time it would take to walk 6 miles at 3 miles per hour: $6 \div 3 = 2$. In other words, it will take 2 hours to walk 6 miles at a speed of 3 miles per hour.

"*Why are we saying all this stuff about distance and speed, while we were supposed to be talking about the Earth, the Moon, and the* Sun?" Joshua-Enoch asked.

"*This is because Baby Earth, Baby Moon, and Baby Sun have also traveled a certain distance at a certain speed before being formed,*" Daddy replied. To figure out how long it took for them to be formed, we need to consider the distance that their babies traveled and at what speed they moved. This is because the Mothers of the celestial bodies traveled from a certain point to another point, meaning over a certain distance, before reaching a position where they were finally wrapped around like spaghetti wrapped around a fork. I hope you remember that we said the Baby celestial bodies were like water layers or spaghetti layers, collected or put together like spaghetti wrapped around a fork.

On top of the time it took for the Mother celestial body to travel the distance I just mentioned, it also took some time for their water layers, or spaghetti layers, or pancake layers to be wrapped around. In the end, the total amount of time it took for the celestial bodies to be formed is the addition of the time their mother traveled and the time it took to wrap their layers of water or whatever materials their babies contained into a round body, looking like a sphere or like spaghetti wrapped around a fork. Now, let's use this math to calculate how long it took for the Earth, the Moon, and the Sun to be formed. I hope you have been enjoying the pretty, artistic pictures we spent a lot of time designing to illustrate the difficult problems we are solving in this book. If yes, let's continue.

19. HOW LONG DID IT TAKE FOR THE EARTH TO BE FORMED?

Using the speed of Mother 2 and the distance separating the Sun and the Earth, my Daddy, Dr. Nathanael-Israel Israel, was the first person in history to figure out how long it took before Baby Earth was born. By dividing the distance separating the Sun and Earth by the speed at which Mother 2 traveled, Daddy calculated how long it took before Mother 2 moved from the position of Baby Sun to about the position where Mother 3 (also called the Mother of Baby Earth and Baby Moon) was born.

Many people across the globe have studied life Earth and the Sun to determine how fast they move, what their size is, and how much distance separates them. For instance, NASA (National Aeronautics and Space Administration) is a big place where astronauts work in the US. As you can see in Figure 24, NASA has shown that the distance between the Sun and the Earth is about 93 million miles (149.6 million kilometers)! Wow! That's huge.

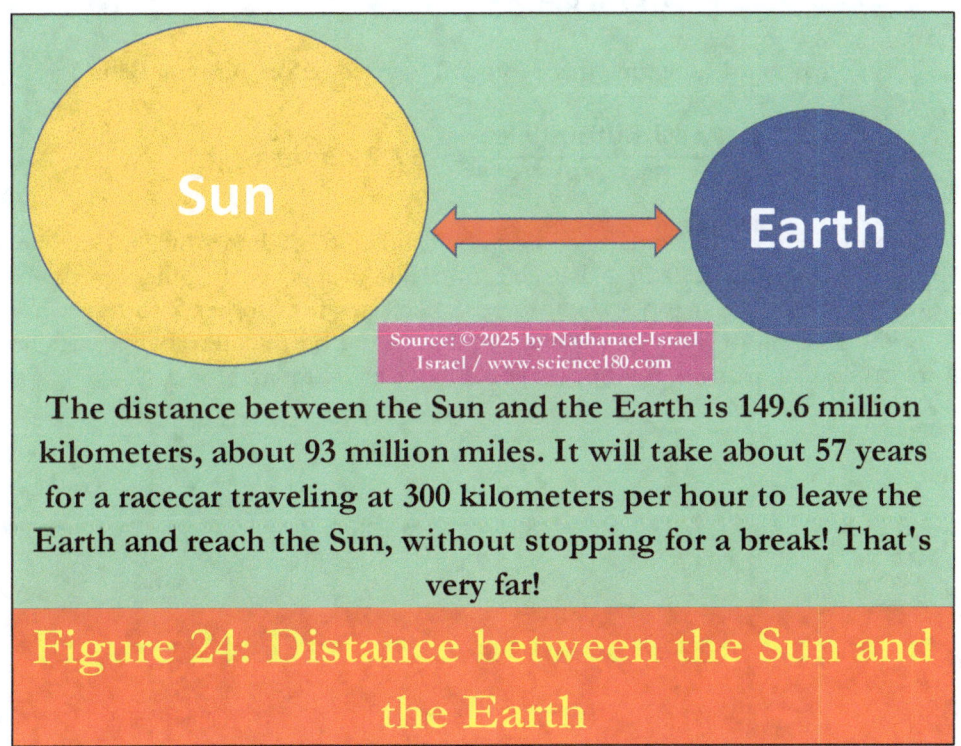

The distance between the Sun and the Earth is 149.6 million kilometers, about 93 million miles. It will take about 57 years for a racecar traveling at 300 kilometers per hour to leave the Earth and reach the Sun, without stopping for a break! That's very far!

Figure 24: Distance between the Sun and the Earth

Based on research using NASA-collected data, Daddy (aka Nathanael-Israel Israel) determined that Mother 2's speed was about 617.6 kilometers per second. That speed of 617.6 kilometers per second is what NASA called "escape velocity" of the Sun. Because Baby Earth was formed quickly after Mother 3 was born, my Daddy proved for the first time in history that, by dividing the distance separating the Sun and the Earth by the speed that Mother 2 traveled, he discovered how long it took before Baby Earth arrived at the position of Earth since the moment Mother 2 started moving away from Baby Sun.

I know how to divide 4 by 2 to get 2. But I don't know how to divide the huge distance between the Sun and the Earth (149.6 million kilometers) by the speed of Mother 2 (617.6 kilometers per second) to obtain the duration I just mentioned. Therefore, to help me, Daddy took his machine called a calculator and plugged the number in to get the result: 149.6 million kilometers divided by 617.6 kilometers per second = 67.29 hours. One more time, the math I just showed you was discovered by my Daddy, and I am very proud of him.

To write down this math in a scientific way, I will first tell you that 149.6 million kilometers is written like 149,600,000 km. Therefore, 149.6 million kilometers divided by 617.6 kilometers per second is scientifically written like:

$$\frac{149,600,000 \text{ km}}{617.6 \text{ km/s}} = 67.29 \text{ hours}$$

In other words, the time that Mother 2 traveled before birthing Baby 3 is:

$$\frac{149,600,000 \text{ km}}{617.6 \text{ km/s}} = 67.29 \text{ hours}$$

Because there are 24 hours in one day, by dividing 67.29 hours by 24, Daddy got 2.8 days, which is more than 2 days, but not quite 3 days yet. In other words, as I showed in Figure 25, it took 67.29 hours or 2.8 days for Mother 2 to travel from about the position of Baby Sun to the position of the Earth where Baby Earth was born. But Baby Earth was not the Adult Earth we know today.

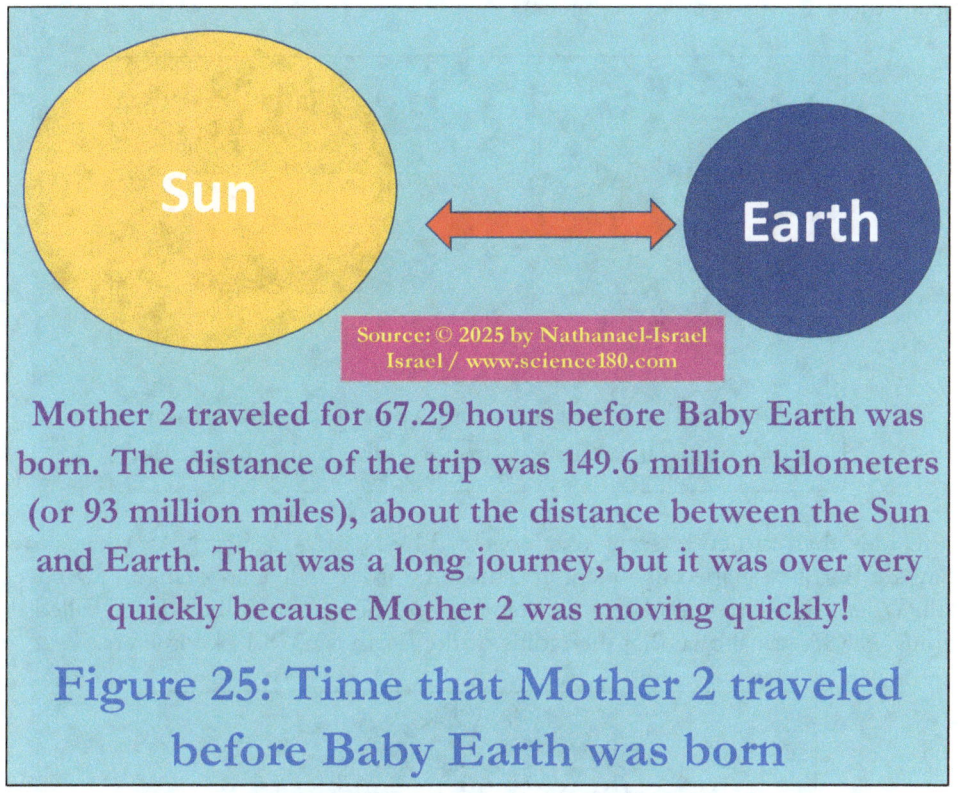

Mother 2 traveled for 67.29 hours before Baby Earth was born. The distance of the trip was 149.6 million kilometers (or 93 million miles), about the distance between the Sun and Earth. That was a long journey, but it was over very quickly because Mother 2 was moving quickly!

Figure 25: Time that Mother 2 traveled before Baby Earth was born

In fact, when Baby Earth was born, it was like water or spaghetti organized in layers or sheets flowing like a river that needed to be rolled around a big fork. My sister Joelle-Major went on to ask Daddy: *"How big were those sheets or layers of water looking like spaghetti that were in Baby Earth?"* *"Great question,"* Daddy replied. Before Daddy answered, I asked him another follow-up question: *"Why was Baby Earth wrapped around like spaghetti, but the dirt or soil on Earth is not sweet like the spaghetti we eat in pasta and rice?"*

"What a fantastic question!" Daddy said. To answer, he first said that when he compared the water layers of Baby Earth to spaghetti, he did not mean that Baby Earth was really spaghetti or noodles. But he meant that the water layers in Baby Earth contained things that were wrapped around like spaghetti or organized in layers like pancakes, one on top of the other. That is why the Adult Earth does not have spaghetti today. However, if we dig deep into the Earth, we will come across soil layers containing spiral structures organized as spaghetti rolled around a fork. For instance, when people dig a well to reach water underground, they discover many layers of soil with different colors and chemicals (see Figure 26). Because some of those spiral things are very small, you may need a microscope to zoom in on them before you can see some of them. Some of those spiral things turned out to be rocks, minerals, atoms, etc. (by the way, "etc." is read as "etcetera," and it

means there are many more examples).

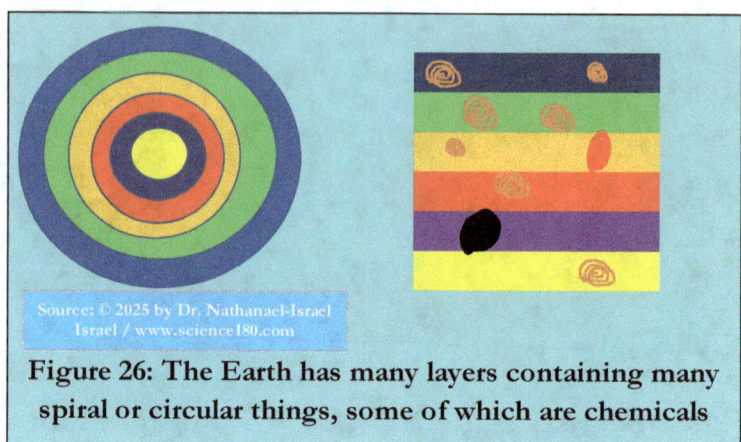

Figure 26: The Earth has many layers containing many spiral or circular things, some of which are chemicals

Daddy then turned to some data collected by NASA. In fact, NASA has shown that the Earth is shaped like a sphere, meaning it looks like an orange or a cheese ball. When things are circular, the distance from the center to the edge is called the radius. NASA has shown that the radius of the Earth is 6378.14 kilometers.

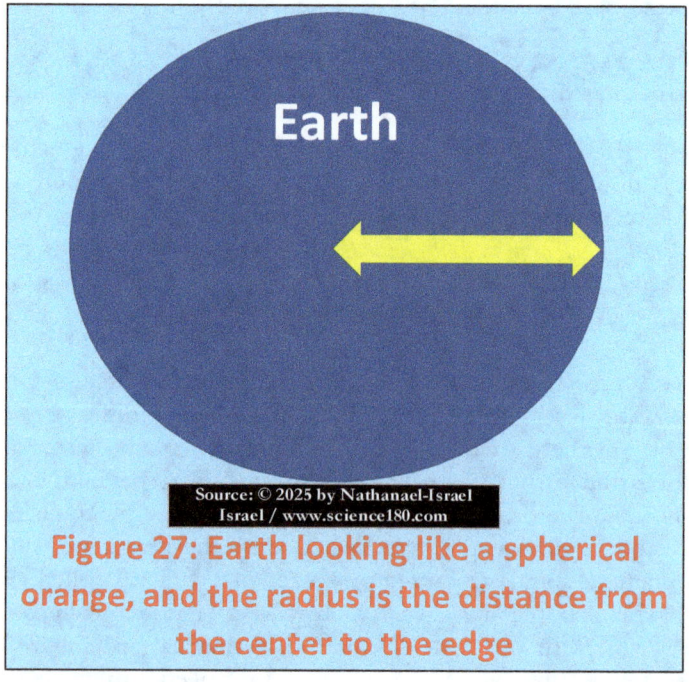

Figure 27: Earth looking like a spherical orange, and the radius is the distance from the center to the edge

Using the radius of the Earth, my Daddy has shown that the length of the sheets

Nathanael-Israel Israel: Acknowledged as Undisputable Specialist of all Questions at the Intersection of Science and Faith

or layers of water or spaghetti in Baby Earth was about the distance around the perimeter of the Earth. That distance all the way around is called the circumference (see Figure 28).

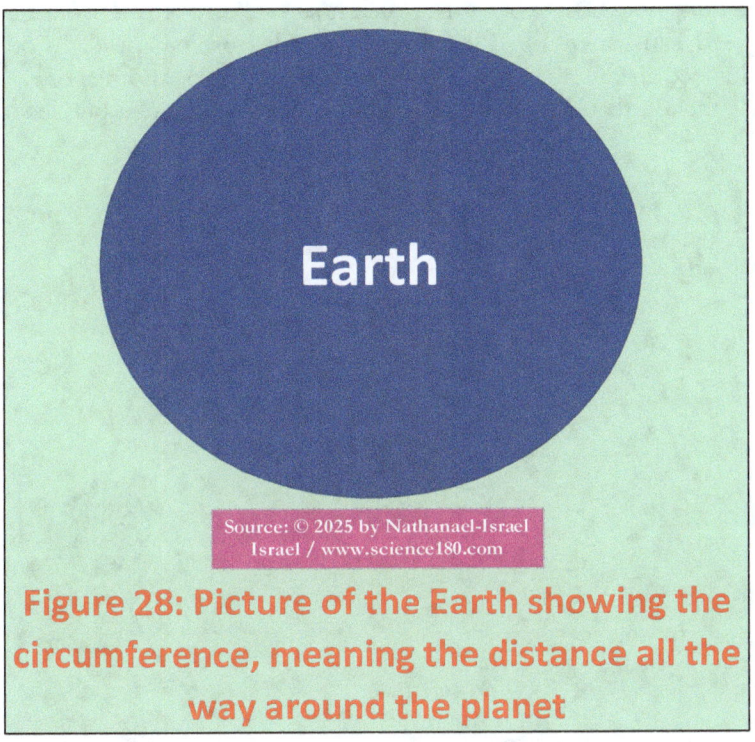

Figure 28: Picture of the Earth showing the circumference, meaning the distance all the way around the planet

If you cannot pronounce that word, don't worry about it; I will do my best to ensure you understand what comes next. Using some math, Daddy calculated the distance to be about 40,054 km. In other words, the water layers in Baby Earth were about 40,054 km. Those water sheets were later wrapped around like 40,054 km of spaghetti rolled around a fork.

Using the math that I did in the story I told you concerning how we walk to school every morning, I hope you remember that time can be calculated by dividing a distance by a speed:

$$\text{Time} = \frac{\text{Distance traveled}}{\text{Speed of the travel}}$$

The math I just showed about time is called a formula. But don't get me wrong: by formula, I did not mean the formula that a mother gives to her baby, like milk, but by formula, I meant a way to write an equation of something. Welcome to our school of STEAM!

Therefore, before calculating how long it took to wrap those water layers, I need to show you the speed at which they were wrapped. In fact, after many years of research, my Daddy showed that the speed at which the water layers in Baby Earth were wrapped around was about the same as the speed at which the Earth moved around the Sun. For those who want to learn some big words, I would like to note that the speed I just mentioned is called the Earth's orbital speed. In other words, the orbital speed means the speed at which a body orbits another one. NASA has shown that the Earth orbits the Sun at about 29.78 kilometers per second (Figure 29).

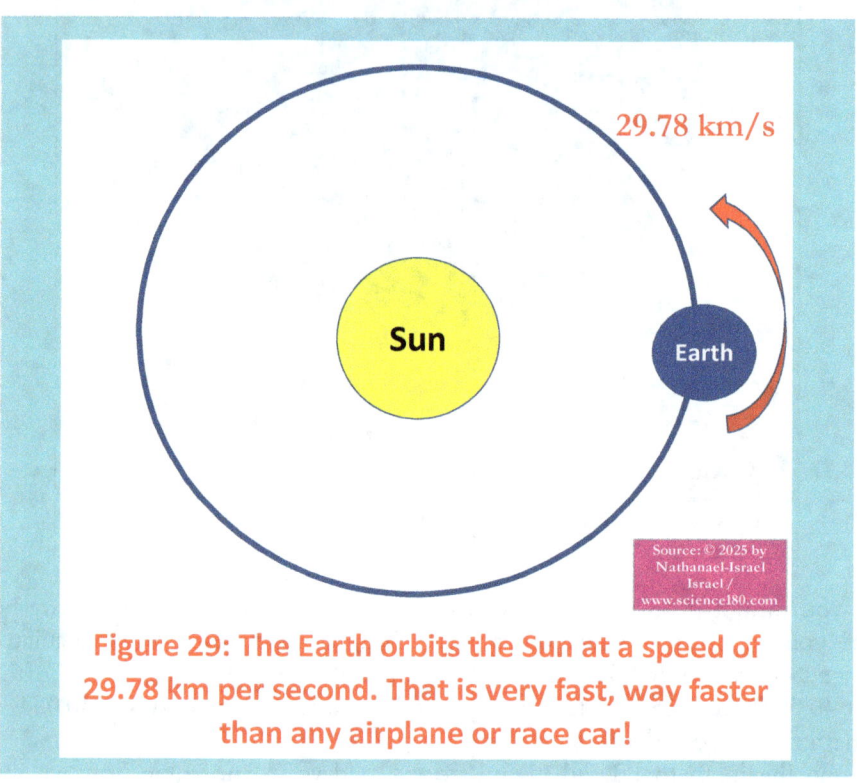

Figure 29: The Earth orbits the Sun at a speed of 29.78 km per second. That is very fast, way faster than any airplane or race car!

Now that we know the distance or length of the water layers or sheets in Baby Earth and also the speed at which they were rolled, we can calculate how long it took for Baby Earth to roll all its water layers around to become Adult Earth. My Daddy was the first person in history to show that by dividing the circumference of the Earth (meaning the distance all around) by the orbital speed (meaning the speed at which the Earth orbits or moves around the Sun), we can know how long it took for Baby Earth to become Adult Earth.

Time to wrap Baby Earth's water layers =
Circumference of the Earth /Orbital speed of the Earth

As a reminder, I told you a while ago that the circumference (meaning the distance all around the Earth) is 40,054 km and the orbital speed (which is the speed of the rolling) is about 29.78 kilometers per second. By dividing 40,054 km by 29.78 km/s, Daddy got 22.41 minutes. Let's put this math in a nice way that scientists will like:

$$\frac{40,054 \text{ km}}{29.7 \text{ km/s}} = 22.41 \text{ minutes}$$

In other words, after Baby Earth was born, it took just 22.41 minutes for its water layers or water sheet that was flowing like a river to be collected into the Adult Earth.

Then, Daddy went on to show us how long it took for the Earth to be fully formed. This amount of time includes 2 things:

- the time it took for Mother 2 to move from about the position of Baby Sun to the position where Baby Earth was born, and
- the time it took for Baby Earth to collect its water layers into a spherical body (looking like an orange) to become Adult Earth.

Remember, I showed you a while ago that after Baby 2 left Baby 1 (meaning Baby Sun), it took 67.29 hours for Baby Earth to be born. By the way, Baby 2 was the Baby that became Mother 2, which was the Mother of all the celestial bodies orbiting the Sun. Daddy demonstrated that, after Baby Earth was born, like water sheets or layers containing long things like spaghetti or noodles, it took about 22.41 minutes to wrap around all its waters to become the Adult Earth we know today. Therefore, according to my Daddy, Dr. Nathanael-Israel Israel, the total time it took for Baby Earth to form and become Adult Earth is 67.29 hours + 22.41 minutes, which equals about 67.66 hours. Because 1 day equals 24 hours, to find how many days there are in 67.66 hours, we divide 67.66 by 24, and we get 2.82 days. Let's put this like a formula:

67.29 hours + 22.41 minutes = 67.66 hours

To convert these hours into days, we divided them by 24:

$$\frac{67.66 \text{ hours}}{24 \text{ hours/day}} = 2.82 \text{ days}$$

By the way, 2.82 days means it wasn't quite the end of the 3rd day yet, but more than 2 days had already passed. In other words, the Earth was born on the 3rd day of creation. In other words, since the beginning of the formation of the Solar System or the universe, it took about 67.29 hours for the Earth to be fully formed. And that was on the third day of creation. All the cool things I am teaching you about how the Universe was formed were discovered by my Daddy, and he taught me those things in a way children my age can understand. Let's recap using some artistic pictures (Figure 30 and Figure 31). We will also use our technology and engineering skills to fix those images.

Baby Earth was like water layers that were rolled like spaghetti around a fork. The water layers were as long as the circumference (perimeter around) of the Earth (which is 40,054 km). The rolling around took 22.41 minutes. In other words, it took 22.41 minutes for the long "noodles" of Baby Earth to become circular "noodles" or the spherical Earth.

After all the water layers were wrapped around, the inside of the Earth looked like layers stacked on top of each other. That is why, when people dig well, they find soil layers.

The Earth looks spherical, but people don't know that its layers are due to how it was formed

Source: © 2025 by Nathanael-Israel Israel / www.science180.com

Figure 30: Water layers of Baby Earth were rolled like spaghetti noodles around a fork

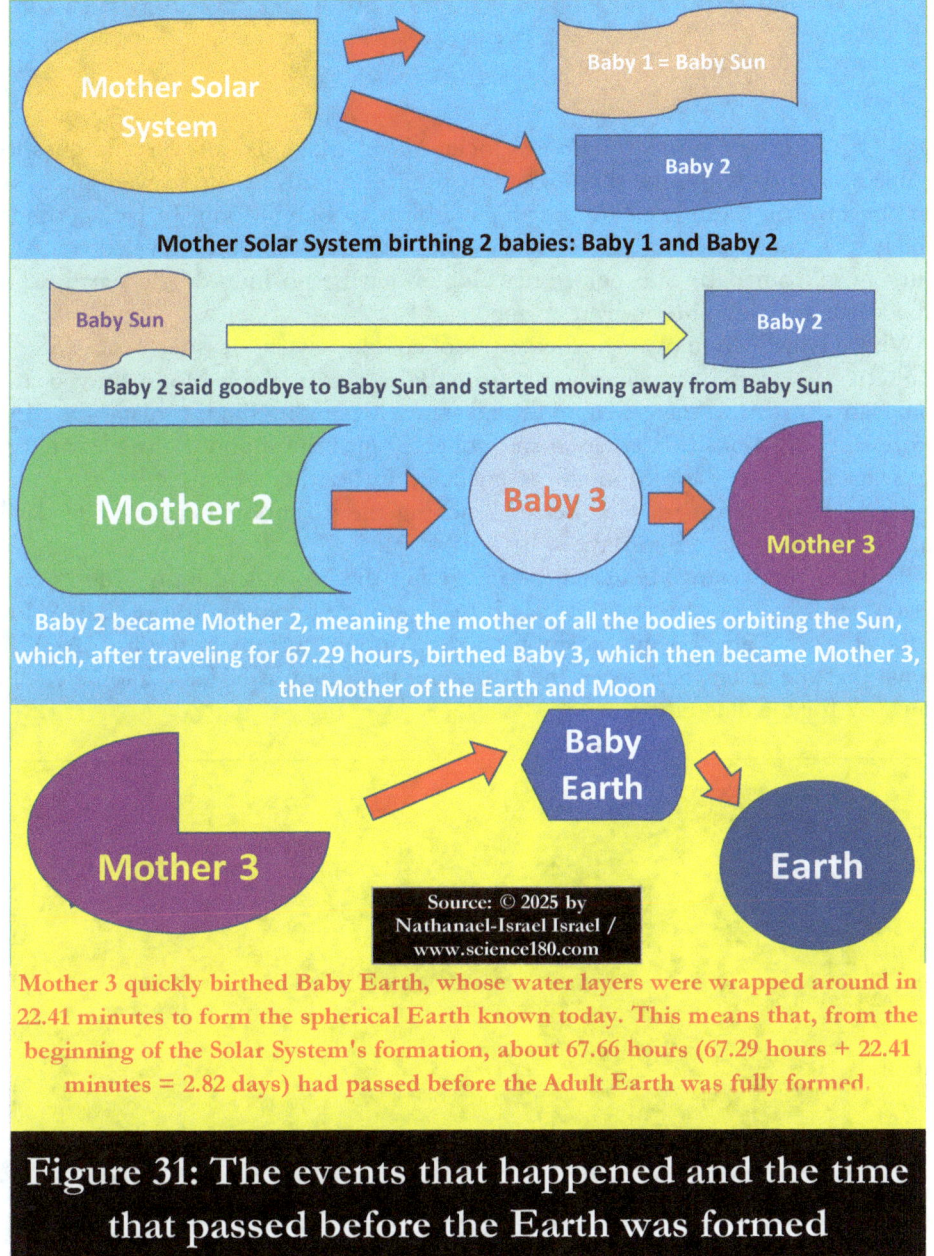

Mother Solar System birthing 2 babies: Baby 1 and Baby 2

Baby 2 said goodbye to Baby Sun and started moving away from Baby Sun

Baby 2 became Mother 2, meaning the mother of all the bodies orbiting the Sun, which, after traveling for 67.29 hours, birthed Baby 3, which then became Mother 3, the Mother of the Earth and Moon

Source: © 2025 by Nathanael-Israel Israel / www.science180.com

Mother 3 quickly birthed Baby Earth, whose water layers were wrapped around in 22.41 minutes to form the spherical Earth known today. This means that, from the beginning of the Solar System's formation, about 67.66 hours (67.29 hours + 22.41 minutes = 2.82 days) had passed before the Adult Earth was fully formed.

Figure 31: The events that happened and the time that passed before the Earth was formed

Now that we are done with the time it took for the Earth to be formed, let's turn to the Moon.

20. HOW LONG DID IT TAKE FOR THE MOON TO BE FORMED?

"*How long did it take to form the Moon*"? Joelle-Major asked. To answer that question, we need to go back to how the Mother of the Earth and the Moon were born, and how, in turn, the Mother of the Earth gave birth to Baby Moon. In fact, Mother 2 (which was the Mother of all celestial bodies orbiting the Sun) traveled for 67.29 hours away from Baby Sun before reaching about the position of the Earth, where Baby Earth and Baby Moon were formed.

When Baby Moon was formed, it looked like layers or sheets of water or spaghetti that started traveling away from Baby Earth. Baby Moon moved away from Baby Earth and traveled about the distance separating the Earth and the Moon before reaching a position where it was collected into the Moon. Before saying how the water layers of Baby Moon were collected into Adult Moon, let's first look at how long Baby Moon traveled before reaching the position where its water layers could be finally collected into the spherical Moon.

By using the distance between the Earth and the Moon and Baby Moon's travel speed, Daddy has done a lot of research to determine how long it took Baby Moon to travel from Baby Earth's position to the Moon's current position. NASA has shown that the distance between the Earth and the Moon is 384,400 kilometers. See Figure 32 for the illustration.

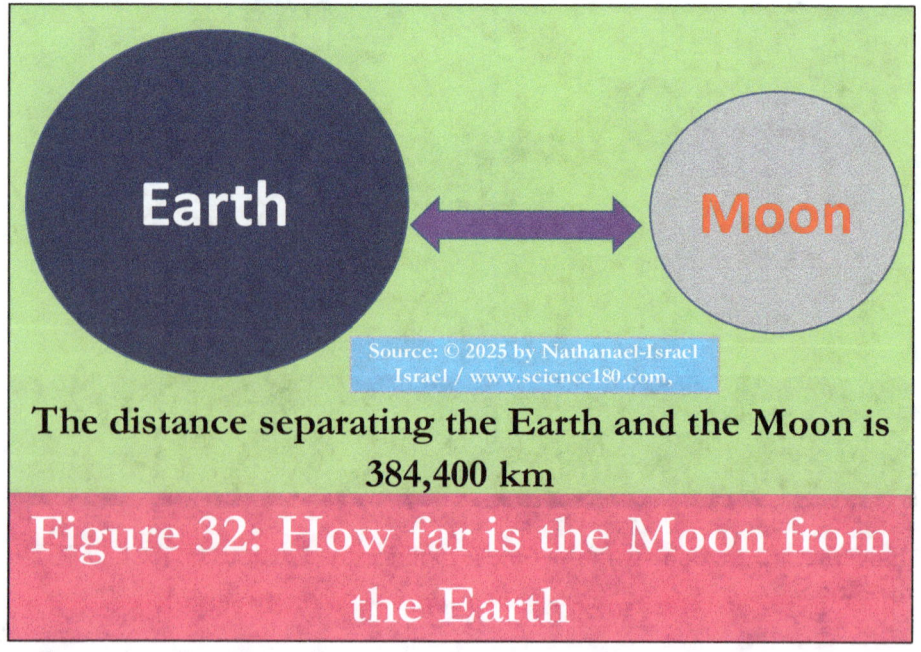

The distance separating the Earth and the Moon is 384,400 km

Figure 32: How far is the Moon from the Earth

SECTION 2: HOW THE GALAXIES, THE PLANETS, THE MOON, AND THE SUN WERE FORMED

Using data collected by NASA on the movement of celestial bodies, my Daddy has shown, for the first time in history, that the speed at which Baby Moon traveled away from Baby Earth was about 11.2 kilometers per second. That speed of 11.2 kilometers per second is called Earth's escape velocity. By the way, the verb "escape" means to leave, run away, or flee from something or someone. For many centuries and decades (a century means 100 years), scientists were aware of escape velocity, but they never knew that Baby Moon escaped Baby Earth at a speed close to Earth's escape velocity. Likewise, for more than 300 years after the work of the great scientist Isaac Newton, scientists did not know that Mother 2 escaped Baby Sun at a speed close to the Sun's escape velocity. My Daddy, Dr. Nathanael-Israel Israel, was the first human being to make such a great discovery. He figured it out after spending about 10 years of research on the origin of the celestial bodies. He was working on this even before any of us, his children, were born.

Therefore, Baby Moon traveled 384,400 kilometers at a speed of 11.2 kilometers per second before reaching the position of the Moon. By dividing the distance traveled by the Moon by the speed of travel, we can figure out how long the journey of Baby Moon took.

Because the distance and the speed of the Moon are too huge, I was not able to do that division by myself. Because my daddy is very smart, he took the calculator again and plugged the number in. When I said Daddy plugged the number into the calculator, I did not mean he plugged it inside the calculator like how we plug a charger into the hole, or an electric car to a battery, but I meant that Daddy used his fingers to enter the number into his calculator:

384,400 kilometers divided by 11.2 kilometers per second, and the answer is 9.53 hours

Written like a formula, what I just said is:

$$\frac{384,400 \text{ km}}{11.2 \text{ km/s}} = 9.53 \text{ hours}$$

I hope you are liking this. As you can see in Figure 33, these 9.53 hours mean that after Baby Moon was formed and left Baby Earth, it took about 9.53 hours before it reached the position of the Moon we know today. By the way, 9.53 hours equals 9 hours, 32 minutes, and 44 seconds.

Science180: All the Universe-Origin and Life-Origin Solutions You Love

When Baby Moon said goodbye to Baby Earth after Mother 3 birthed both, Baby Earth asked Baby Moon, "Where are you going, my friend? I like you, don't leave me".

Mother 3

Baby Earth

Baby Moon

Baby Moon replied to Baby Earth, "Don't worry, I will not be that far, I am just traveling for 9.53 hours to go to my own place. You can still see me anytime. Did you forget that when you were being born, I was pushed away, and that for both of us to exist, there is no way we can stay together"?

Source: © 2025 by Nathanael-Israel Israel / www.science180.com

Earth

Moon

After saying goodbye to Baby Earth, Baby Moon traveled for 9.53 hours before crossing the 384,400 km and reaching the position of the Moon, where it took some additional time before being wrapped around into the Moon.

Figure 33: How long did Baby Moon travel after leaving Baby Earth before reaching the position of the Moon today?

When Baby Moon reached the position of the Moon, it was still looking like layers of water or spaghetti or noodles that needed to be wrapped around a fork so that Baby Moon could become the Adult Moon we know today. The length of the water layers of Baby Moon was as long as the perimeter or circumference of the Moon we know today. If you remember, I told you earlier that the distance from the center of a circle to the edge is called the "radius," and that using the radius, the perimeter, or circumference, can be calculated.

NASA has determined that the Moon's radius is 1738.1 kilometers. Using that number, my Daddy said the perimeter, or the distance around the Moon, is about 10,915.27 kilometers (see Figure 34).

Nathanael-Israel Israel: Acknowledged as Undisputable Specialist of all Questions at the Intersection of Science and Faith

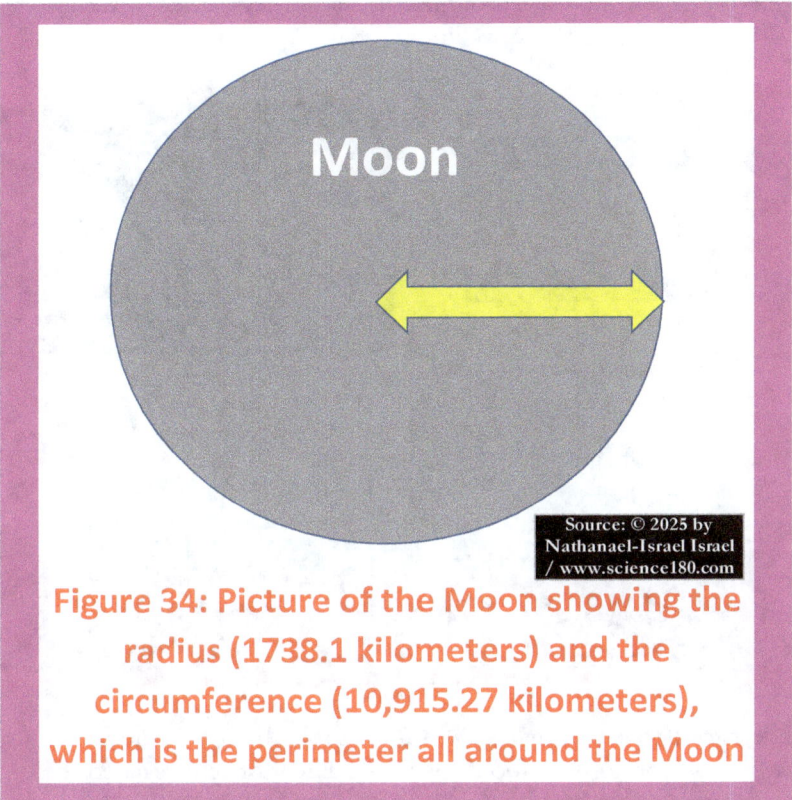

Figure 34: Picture of the Moon showing the radius (1738.1 kilometers) and the circumference (10,915.27 kilometers), which is the perimeter all around the Moon

Recalling the circumference (meaning the length of the perimeter of the Earth), Daddy said that the length of the water layers of the Baby Moon was about one-quarter that of the Earth. This means that the water layers of Baby Earth were about 4 times those of Baby Moon.

Now that we know the length of the water layers of Baby Moon that needed to be wrapped around like a fork, we still need to know the speed of that wrapping around. Again, based on some data collected by NASA, my Daddy has shown for the first time in history that the speed at which the water layers of Baby Moon were rolled around was about 1.022 kilometers per second (which is about 1 kilometer per second). That speed is called the orbital speed of the Moon.

As we did before, to know how long it took for the water layers of Baby Moon to be wrapped around, my Daddy divided the circumference of the Moon by the orbital speed of the Moon:

$$\frac{10,915.27 \text{ km}}{1.022 \text{ km/s}} = 2.97 \text{ hours}$$

In other words, as you can see in Figure 35, after Baby Moon arrived at the position of the Moon, it took around 2.97 hours for all its water layers to wrap around to form the Adult Moon we know today.

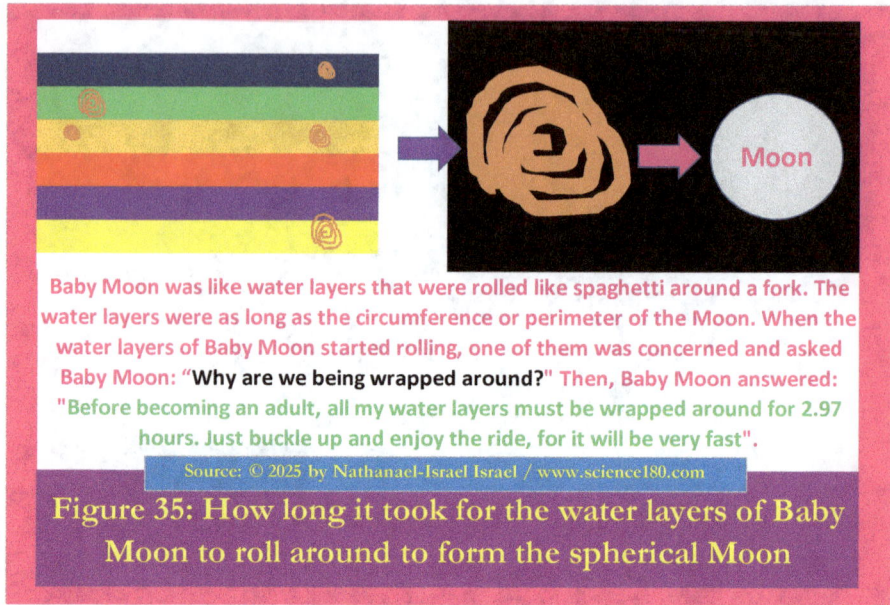

Baby Moon was like water layers that were rolled like spaghetti around a fork. The water layers were as long as the circumference or perimeter of the Moon. When the water layers of Baby Moon started rolling, one of them was concerned and asked Baby Moon: "**Why are we being wrapped around?**" Then, Baby Moon answered: "Before becoming an adult, all my water layers must be wrapped around for 2.97 hours. Just buckle up and enjoy the ride, for it will be very fast".

Source: © 2025 by Nathanael-Israel Israel / www.science180.com

Figure 35: How long it took for the water layers of Baby Moon to roll around to form the spherical Moon

A while ago, I told you that Baby Moon took about 9.53 hours to move from Baby Earth's position to the Moon's position. Then, once it reached the position of the Moon, it took about 2.97 hours for all its water layers to be wrapped around to form the Adult Moon we all know today. This means that the time it took for Baby Moon since it left Baby Earth, all the way to the time that Adult Moon was formed, was equal to:

$$9.53 \text{ hours} + 2.97 \text{ hours} = 12.5 \text{ hours}$$

In other words, after Mother 2 birthed Baby 3, which became Mother 3, meaning the Mother of the Earth and the Moon, it took about 12.5 hours before the Moon was fully formed as an adult.

I showed you earlier that after leaving Baby Sun, Mother 2 spent about 67.29 hours before giving birth to the Mother of the Earth and Moon. I just showed you that after Baby Moon was born, it took 12.5 hours for it to grow up and become the adult Moon we know today. To calculate the total amount of time that passed before the Moon was formed since the beginning, we need to add 67.29 hours to 12.5 hours. The result of this addition is about 79.8 hours. Because one day is 24 hours, 79.8 hours is equal to 3.32 days. Let's do some STEAM and lay out this sentence like a beautiful math formula:

Nathanael-Israel Israel: Acknowledged as Undisputable Specialist of all Questions at the Intersection of Science and Faith

$$\frac{79.8\ \text{hours}}{24\ \text{hours/day}} = 3.32\ \text{days}$$

In other words, the Moon was fully formed about 3.32 days after the beginning of the Solar System. By the way, 3.32 days means 3 days have passed, but 4 days have not yet passed. Therefore, 3.32 days means the 4th day. That is why scientific data show that the Moon formed on the 4th day after the beginning of the Solar System's formation. Figure 36 summarizes the process and the time it took for the Moon to be formed.

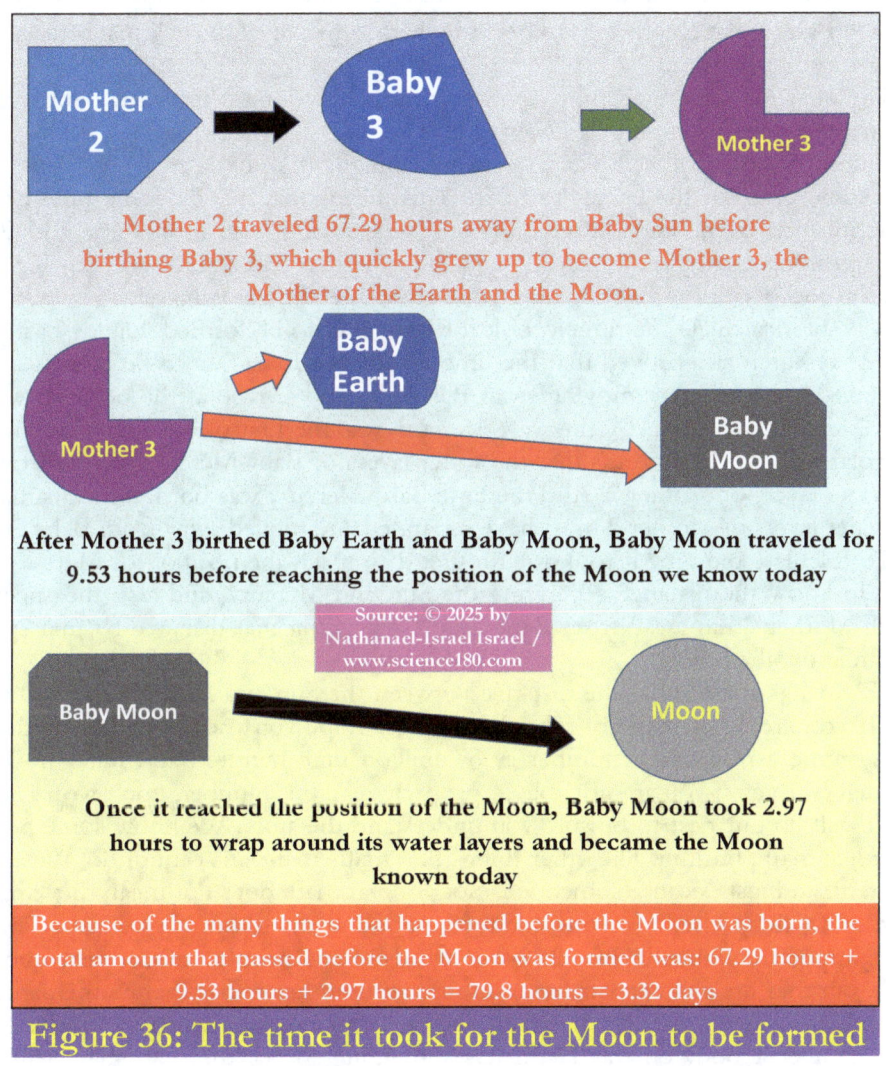

Figure 36: The time it took for the Moon to be formed

21. HOW LONG DID IT TAKE FOR THE SUN TO BE FORMED?

I then asked Daddy, "How long did it take for the Sun to be born?"

Using the same math he did for the Earth and the Moon, Daddy helped me answer that question. As a reminder, we learned earlier that Mother Solar System gave birth to 2 babies:

- Baby 1 or Baby Sun, which grew up to become the Sun, and
- Baby 2, which grew up to become the Mother of all the celestial bodies orbiting the Sun.

The question we are trying to answer now is how long it took for Baby Sun to become the fully Adult Sun we know today.

Indeed, when Mother Solar System was pregnant with Baby 1 and Baby 2, it took some time for the babies to form. This means that it took some time before Baby Sun was born. Then, after Baby Sun was born, it also took some additional time (meaning more time in addition to the time that had already passed) for it to grow up and become Adult Sun. Are you getting me? I hope you are!

Let's first calculate the time it took for Baby Sun to be formed. Daddy has done a lot of research and proved that the time it took for Baby Sun to be born is about the time Mother 2 spent moving from Baby Sun and arrived at the position where Baby Mercury was born. By the way, Mercury was the first planet that was born in the Solar System. This means that the water layers of Baby Mercury were on top of the water layers of Mother 2. And the time Baby Mercury was born, it means that all the water layers of Mother 2 must have separated and moved away from Baby Sun.

To calculate the time it took for Mother 2 to reach the position of Mercury, we need to know the distance separating the Sun and Mercury, and also the speed at which Mother 2 ran away from Baby Sun. By dividing that distance by that speed, we can know the time.

NASA has shown that the distance between the Sun and Mercury is 57,909,000 km. If you are an elementary school student, I hope you tried your best and read this gigantic astronomical number as 57 million nine hundred and nine thousand kilometers. Even if you are not able to properly read this number, don't worry about it. We will do our best to be sure you understand the story. We know some people don't like math, but they like other things that mathematicians cannot do. We are all different, and just because someone is not good at math does not mean they are not good at anything else. Someone cannot be good at math, but very good at art, or other important things in life. In other words, some people are not very good at math, but they are still very smart, and they can do other things that mathematicians and other scientists cannot do. We need to learn how to humble ourselves and know that everybody on this Earth has something special they can share with the rest of the world! This special thing can be a great work of art, a great cultural thing, a great musical thing, a great technological thing, a great piece of handwork, and many more. Everybody is unique, and we don't need to keep comparing ourselves to others. Also, some people don't want to hear about math or science, yet they are enjoying our story. We enjoyed Daddy teaching us about science and life in general!

Daddy has demonstrated that Mother 2's speed of travel was about 617.6 kilometers per second. Before we continue, let me teach you some great scientific words. The 57,909,000 km separating the Sun and Mercury is called the semi-major axis of Mercury, which means the average distance separating Mercury and the Sun. The 617.6 km/s is what scientists have called the escape velocity of the Sun, which means the speed that things at the surface of the Sun must have before they can be able to escape or run away from the Sun without the gravity of the Sun pulling them back. In fact, just as the gravity of the Earth pulls things toward the Earth, so also the gravity of the Sun pulls things toward the Sun.

A long time ago, even before gravity formed, Baby Sun could also pull things toward its surface. This means that Baby Sun could have held Baby 2 close to it if Baby 2 was very slow, like a tortoise or a sloth. But, because Baby 2 was able to run away from Baby 1, Daddy had done some great math that showed that Baby 2 escaped or fled Baby Sun with the speed almost equal to the escape velocity of the Sun (617.6 kilometers per second). What I am saying here may sound silly or funny, but trust me, it took 10 years for my Daddy to dig deep into this crazy stuff until he got to the bottom of this difficult problem of the formation of the WHOLE universe. Did I tell you that my Daddy obtained his PhD or doctorate (a PhD is like grade 24) in science in the USA? For now, let's not worry about that, but get back to our math to know the time that has passed before Baby Mercury was formed.

Remember, we said that to calculate that time, we need to consider the distance separating the Sun and Mercury and also the speed with which Baby 2, which became Mother 2, fled Baby Sun. Here, the distance is 57,909,000 kilometers, and the speed is 617.6 kilometers per second. Therefore, the division that we need to do here is:

$$\frac{57,909,000 \text{ km}}{617.6 \text{ km/s}} = 26.05 \text{ hours}$$

In other words, my Daddy was the first person in the whole world to ever show that, since the beginning of creation, it took about 26.05 hours before Baby Sun was born. But this was not the Adult Sun we have today yet, for some changes happened to Baby Sun before it became the Adult Sun.

In fact, after Baby Sun was born, it looked like layers of fiery (meaning that they have fire) materials containing water or plasma organized like pancakes in a flowing river consisting of fire or burning stuff. Some of that stuff included water. In other words, although it is very hot, the Sun also contains water as of today. The layers of materials in Baby Sun needed to be gathered together before the Sun could be formed. Using the radius of the Sun, Daddy was able to calculate the length of the material layers of Baby Sun.

NASA has measured the Sun's radius and found it to be 695,700 kilometers. Doing some math, Daddy said that this radius implies that the circumference or perimeter or the distance all around the Sun is 4,368,996 kilometers (Figure 37). In other words, the layers of the materials in Baby Sun were about four million three hundred sixty-eight thousand, nine hundred, ninety-six kilometers long. Those layers of stuff were like spaghetti that needed to be wrapped around a fork.

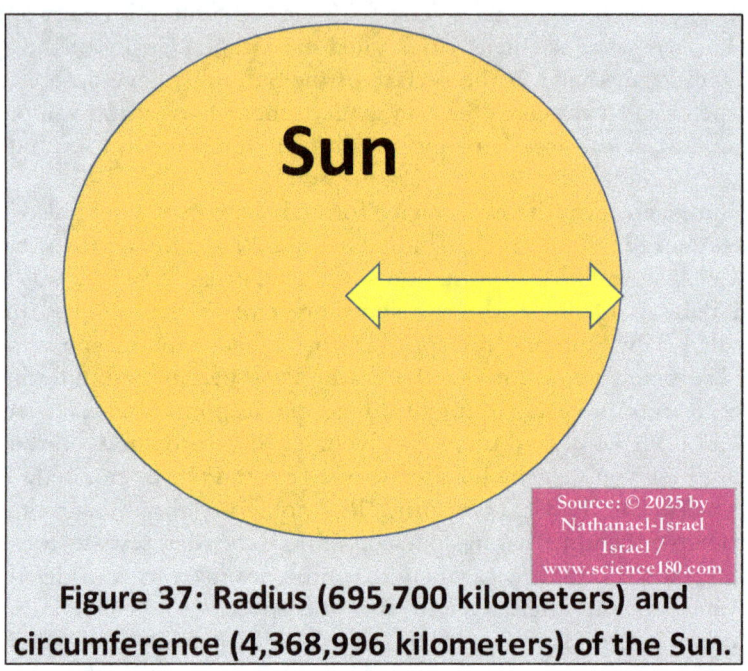

Source: © 2025 by Nathanael-Israel Israel / www.science180.com

Figure 37: Radius (695,700 kilometers) and circumference (4,368,996 kilometers) of the Sun.

To calculate how long it took for these layers to be wrapped around, we need to know the speed of that wrapping. Using some data collected by NASA, Daddy showed that the speed of the wrapping around was about 19.4 kilometers per second. That speed is also known as the Sun's speed relative to nearby stars. Like we did before for the Earth and the Moon, time is a distance divided by a speed. When Daddy divided 4,368,996 kilometers by 19.4 kilometers per second, he got 62.56 hours.

$$\frac{4,368,996 \text{ km}}{19.4 \text{ km/s}} = 62.56 \text{ hours}$$

In other words, after Baby Sun was born, it took about 62.56 hours for it to wrap around all its layers of material or plasma to become the Adult Sun as we know it today.

Nathanael-Israel Israel: Acknowledged as Undisputable Specialist of all Questions at the Intersection of Science and Faith

SECTION 2: HOW THE GALAXIES, THE PLANETS, THE MOON, AND THE SUN WERE FORMED

To calculate the total amount of time it took for the Sun to be fully formed, we need to add 26.05 hours (which was the time it took for Baby Sun to be formed) and 62.56 hours (which was the time it took for Baby Sun to mature or grow up to become the Adult Sun). The addition of these two times gives 88.6 hours:

$$26.05 \text{ hours} + 62.56 \text{ hours} = 88.6 \text{ hours}$$

This means that, from the beginning, it took 88.6 hours for the Sun to fully form. Because one day is equal to 24 hours, to know how many days are in 88.6 hours, Daddy divided 88.6 hours by 24, and the answer is 3.69 days:

$$\frac{88.6 \text{ hours}}{24 \text{ hours/day}} = 3.69 \text{ days}$$

By doing this math, Daddy showed that 3.69 days after the beginning of creation, the Sun was fully formed. The number 3.69 days means that 3 days had passed, but the 4th day was not finished yet. In other words, 3.69 days means that the Sun was formed on the 4th day. Those who finished or who are about to finish elementary school know that 3.69 days is a decimal that lies between 3 and 4, meaning a number that has a comma in it and which is greater than 3 but smaller than 4. All decimals have a whole and a fractional part. For 3.69 days, the whole is 3, and the fractional part is 0.69. In other words, 3.69 days = 3 days + 0.69 days. Hence, as you can see in Figure 38, the Sun was formed on the 4th day.

Mother Solar System

Baby Sun

Baby 2

After Mother Solar System birthed its 2 children, it took about 26.05 hours before Baby 2 left Baby Sun (which was Baby 1). At that time, Baby Sun did not yet have a clear shape, but was a layer of plasma or other hot material that needed to be collected to form the Sun. Baby 2 would later grow up to become Mother 2, the Mother of all the bodies orbiting the Sun

Baby Sun

Sun

After Baby 2 completely left Baby Sun, it took 62.56 hours for the plasma layers of Baby Sun to collect and form the spherical Sun.

Source: © 2025 by Nathanael-Israel Israel / www.science180.com

The total amount of time it took for the Sun to be formed was 26.05 hours + 62.56 hours = 88.6 hours, which is 3.69 days, meaning on the 4th day Since the beginning of the formation of the Solar System

Figure 38: How long it took for the Sun to be formed

I hope you like our STEAM explanation. Before we do anything else in this book, we will check whether the Earth, the Moon, and the Sun were really formed as the Bible says. This is a BIG deal!

22. DOES THE MATH WE DID IN THIS BOOK MATCH THE BIBLICAL STORY OF CREATION?

Before we say whether the story and math in this book match the Biblical story of creation, we will first review what our math says. Then, we will compare it with what the Bible said about creation.

Based on the math we did using the scientific data that NASA has collected and that my Daddy, Dr. Nathanael-Israel Israel, has better analyzed, the Earth was fully formed on the 3rd day of creation. In contrast, the Moon and the Sun were fully formed on the 4th day (see Figure 39). During their formation, water layers were separated, transported over long distances, and then gathered together to form round celestial bodies such as planets, asteroids, the Sun, and other stars in the universe.

Now, let's look at what the Bible said about the creation of the Earth, the Moon, and the Sun.

In fact, the Bible said that the formation of the Earth was completed on the 3rd day of creation. Before that, the Bible's Book of Genesis (chapter 1:6-13) said that there was a time when Baby Earth was filled with water, and did not have a form. Then the Bible said that the waters were gathered together, and at one point some covered the surface of the land. Then, God moved the waters on the surface of the water so they could flow into the oceans, seas, rivers, ponds, lakes, creeks, and other bodies of water. All this happened on the 3rd day when the creation of the Earth was finished. In other words, the math we did using NASA data perfectly matched the Bible's account of the creation of the Earth. In other words, the math we did using scientific data and the Biblical story of creation both indicated that the formation of the Earth was completed on the 3rd day. This means the Biblical account of Earth's formation is true.

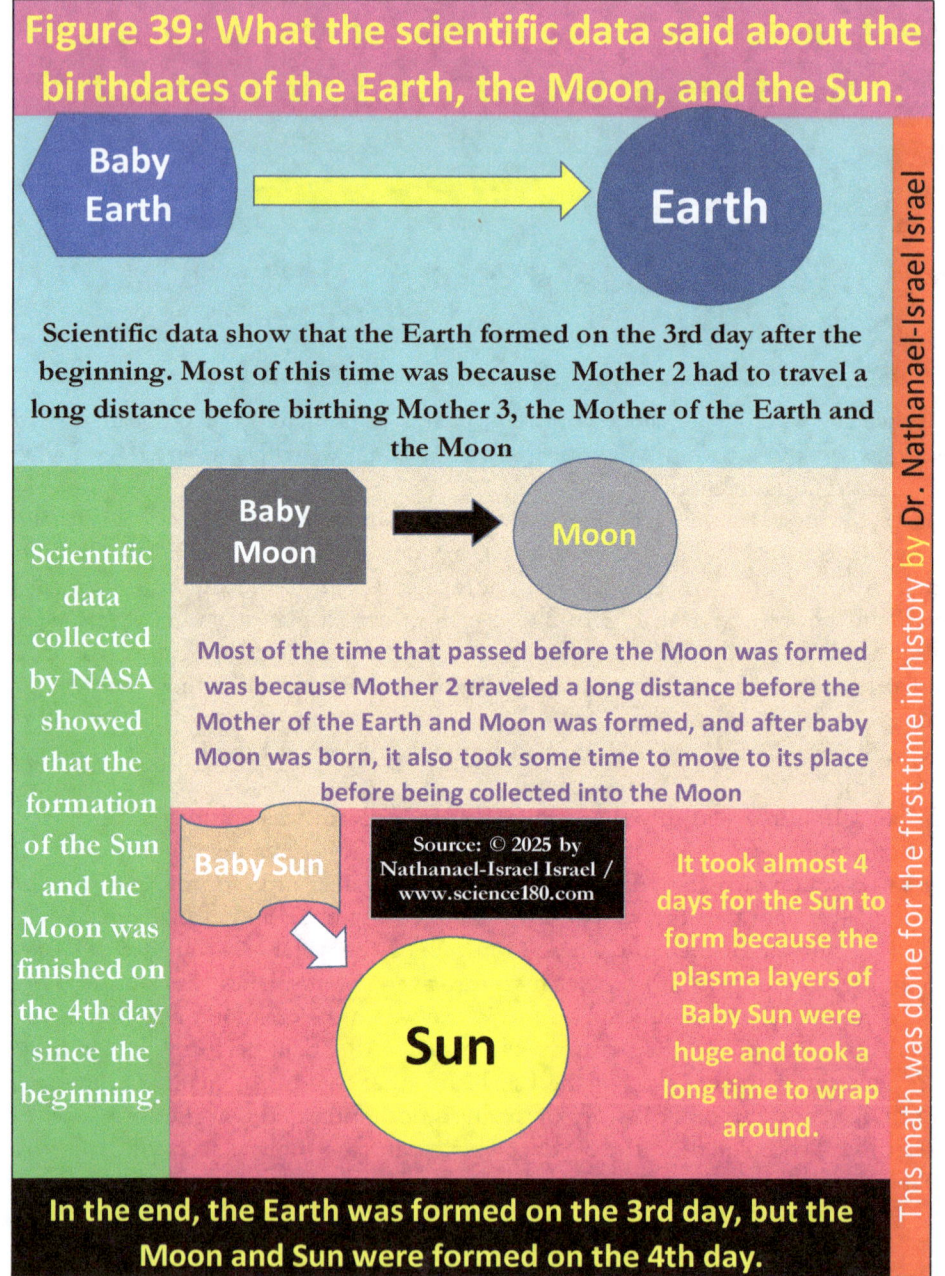

Figure 39: What the scientific data said about the birthdates of the Earth, the Moon, and the Sun.

Baby Earth → Earth

Scientific data show that the Earth formed on the 3rd day after the beginning. Most of this time was because Mother 2 had to travel a long distance before birthing Mother 3, the Mother of the Earth and the Moon

Baby Moon → Moon

Scientific data collected by NASA showed that the formation of the Sun and the Moon was finished on the 4th day since the beginning.

Most of the time that passed before the Moon was formed was because Mother 2 traveled a long distance before the Mother of the Earth and Moon was formed, and after baby Moon was born, it also took some time to move to its place before being collected into the Moon

Baby Sun

Source: © 2025 by Nathanael-Israel Israel / www.science180.com

It took almost 4 days for the Sun to form because the plasma layers of Baby Sun were huge and took a long time to wrap around.

Sun

This math was done for the first time in history by Dr. Nathanael-Israel Israel

In the end, the Earth was formed on the 3rd day, but the Moon and Sun were formed on the 4th day.

Now, let's see how the math we did for the Moon aligns with or contrasts with the Biblical account of the Moon's formation. My Daddy showed in a previous chapter that the Moon was formed 3.32 days after the beginning, meaning on the 4th

Nathanael-Israel Israel: Acknowledged as Undisputable Specialist of all Questions at the Intersection of Science and Faith

day of creation. This is exactly what the Bible says in Genesis 1:14-19. In other words, the formation of the Moon, as shown using the scientific data, is exactly what the Bible said.

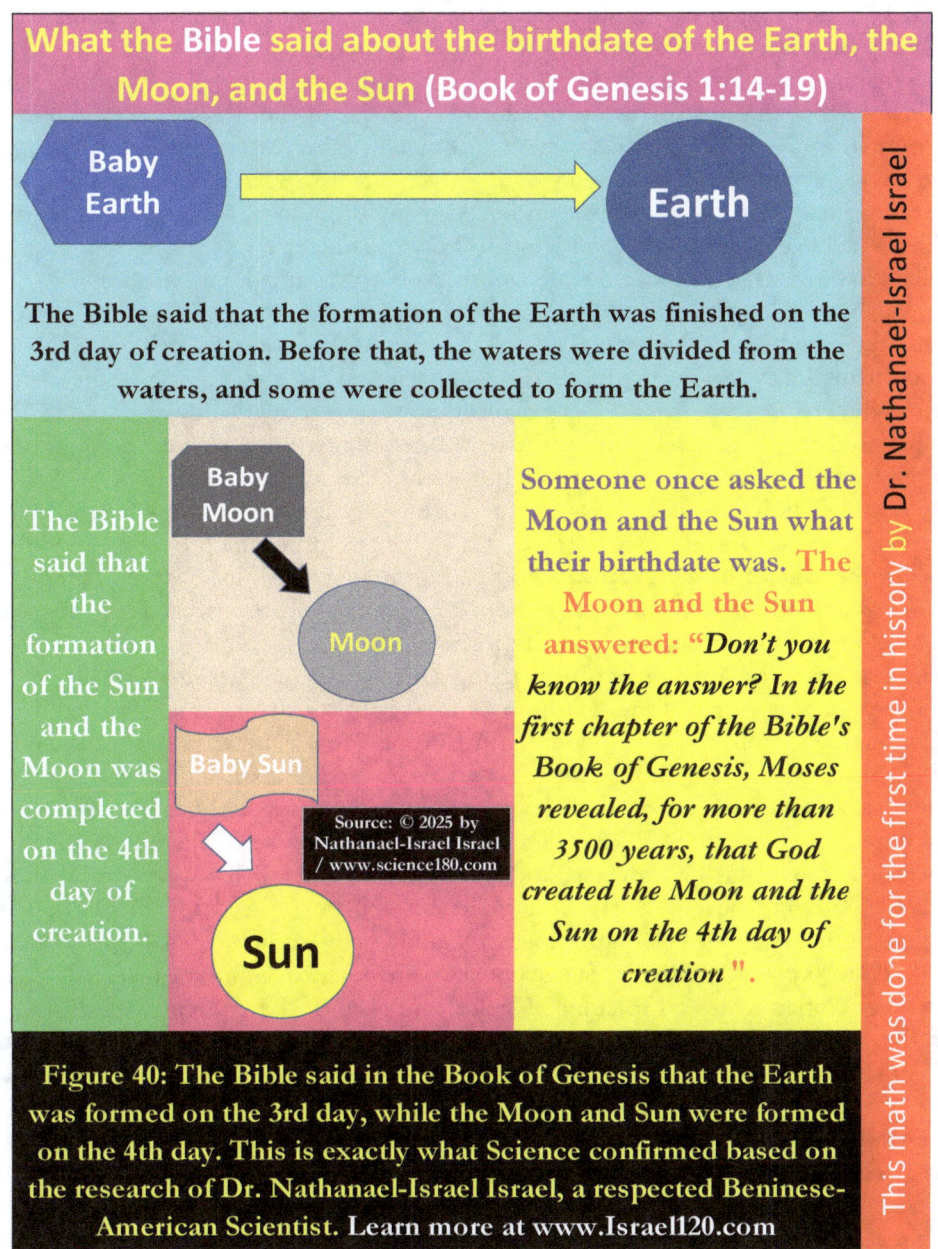

What the Bible said about the birthdate of the Earth, the Moon, and the Sun (Book of Genesis 1:14-19)

Baby Earth

Earth

The Bible said that the formation of the Earth was finished on the 3rd day of creation. Before that, the waters were divided from the waters, and some were collected to form the Earth.

Baby Moon

Moon

Baby Sun

Source: © 2025 by Nathanael-Israel Israel / www.science180.com

Sun

The Bible said that the formation of the Sun and the Moon was completed on the 4th day of creation.

Someone once asked the Moon and the Sun what their birthdate was. The Moon and the Sun answered: *"Don't you know the answer? In the first chapter of the Bible's Book of Genesis, Moses revealed, for more than 3500 years, that God created the Moon and the Sun on the 4th day of creation"*.

This math was done for the first time in history by Dr. Nathanael-Israel Israel

Figure 40: The Bible said in the Book of Genesis that the Earth was formed on the 3rd day, while the Moon and Sun were formed on the 4th day. This is exactly what Science confirmed based on the research of Dr. Nathanael-Israel Israel, a respected Beninese-American Scientist. Learn more at www.Israel120.com

Lastly, let's compare the Biblical account of the Sun's creation with the scientific demonstration my Daddy did. As a reminder, I showed you in a previous chapter that the Sun was formed 3.69 days after the beginning of creation, meaning on the 4th day. This date is exactly when the Bible said the Sun was formed. One more time, the Biblical account of the Sun's formation perfectly matches the scientific data. In other words, the creation story of the Sun is true. Figure 40 summarizes how the Biblical account of creation matches the scientific data.

To summarize, my Daddy, Dr. Nathanael-Israel Israel, showed for the first time in history that the date of the formation of the Earth, the Moon, and the Sun as recounted in the Bible is 100% correct and matches the date calculated using math and the data that top scientists, including those at NASA, have collected over the years. In fact, for many hundreds, and even many thousands, of years, people from all nations have been trying to understand how the universe was formed. Billions of dollars have been spent, but no real answer has been obtained until now.

All these facts prove that the Bible is correct and that the God of the Bible is really the Creator. By the way, the Biblical story of creation was written by Moses more than 3500 years ago. In those days, science did not exist, yet the details revealed by Moses were very accurate, meaning very precise. Daddy also told us that he has studied all other religions in the world, but none of them have a clear story of creation of the universe that matches the scientific evidence. In other words, of all the religious and scientific books in the world, only the Bible contains an original story of how the Universe was formed, just as the scientific data proves.

For the sake of history, Daddy said that before he closed this chapter, he needed to tell us something very important about some great scientists who lived before us and who have tried to understand the formation of the universe. Indeed, for more than 300 years, scientists have been trying very hard to get to the bottom of the problem of our origin. One of the greatest scientists of all time is called Isaac Newton. He is from the UK, a country in Europe. He was born in 1643 and died in 1727, meaning he died about 300 years ago. He tried to explain the origin of the universe, and he did a lot of work on gravity. But his efforts did not prove much in the universe; he was thinking the right way. Then another scientist, Albert Einstein, also tried to explain the universe, and he is best known for his work on relativity, which is not the business of a little child, so you don't need to worry about it. Although many people celebrate Einstein, Daddy said that he made a bigger mistake than Isaac Newton. Other great scientists were Galileo Galilei, Leonhard Euler, Nicolaus Copernicus, and Johannes Kepler, and each of these scientists told a story of the universe as they thought it could have happened; those stories are what scientists call theories. But none of them has demonstrated what my Daddy did. Otherwise, they could have explained how the Earth, the Moon, the Sun, and everything else in the universe were formed. In other words, until the research of my Daddy, Dr. Nathanael-Israel Israel, no human being has ever explained the formation of the universe properly. Some people even thought that the Bible was lying. Yet its story is true, but it is people who failed to explain it.

"How do you feel about the story and the fact that the Bible story of creation matches the

Nathanael-Israel Israel: Acknowledged as Undisputable Specialist of all Questions at the Intersection of Science and Faith

scientific data"? That was the question Daddy asked us as he came to this point. We all said that we feel very good and happy about the story. We also told Daddy that we are very proud of him. We thank God because we (Josephine Israel, Joelle-Major Israel, and Joshua-Enoch Israel) are the first children in the whole world to learn about this story and even to help our Daddy (Dr. Nathanael-Israel Israel) write a book about it!

I hope you enjoy the cool pictures we took. I hope you will agree with the countless people who have acknowledged that my dad, Nathanael-Israel Israel, is the first human being to reconcile science with the Biblical account of creation.

The math and the story in this book gave us more confidence that the God of Israel described in the Bible is truly God. I am glad I know Him and have believed in Him all my life. The proof of the universe's creation presented in this book also encouraged me to believe in Him forever and ever.

What I shared with you in this book is just a small portion of what my dad discovered, and it is something children ages 7-12 could fully understand. In fact, Daddy has written many other books on this topic, and if you want to know more, please consult his website at www.Science180.com or www.Israel120.com. Because Daddy has so many funny stories, I think he should write other books for us about life and chemicals, just like he did for adults.

By the way, we will still have a lot of great questions coming up. If you want to know the whole story, please keep reading. You don't want to miss what is coming up! Are you ready to learn more? Let's keep going!

23. IS THERE ANY OTHER COOL SCIENTIFIC THING YOU CAN LEARN ABOUT THE CREATION OF THE UNIVERSE AND GOD?

As you can see from my Daddy's demonstration, the scientific data points perfectly to God as the creator of the universe, just as the Bible says. Although I have a lot to say about God and the creation of the universe, I cannot do so in this book, for I don't want to force people who don't believe in God to read it if they don't want to. At the same time, we cannot reach this point in this book without bringing to your attention another children's book my dad wrote about the creation of the universe, in which he addresses some deep questions children and adults usually ask about God and the Biblical creation. Even if you don't believe in God, I think that his book ("*How God Created Baby Universe*") will greatly benefit you. In his other books, my dad addressed other important questions, such as:

- Is it true that God created everything?
- Who and where is God?
- Who made God?
- What does He look like?
- What does heaven look like?
- When God spoke for things to appear during creation, like the Bible says, where did they come from?
- When the animals were created, did they obey God's instructions?
- Where did God live before He created the heavens and the Earth?
- Where was God before creation?
- Why can't we live forever like God?
- How come everybody does not believe in God?
- How did God make heaven?
- How did God make Himself, and what material is He made from?
- How did God make the galaxies?
- Why did God take six days to create everything?
- Why does God live forever, but we, human beings, can't—or can we?
- How old is God?

Nathanael-Israel Israel: Author of "Reconciling Science and Creation Accurately"

SECTION 2: HOW THE GALAXIES, THE PLANETS, THE MOON, AND THE SUN WERE FORMED

- How old is the Earth?
- Before God created the first human beings, what was the universe and God himself like?
- Did God create any bad animals that can hurt people?
- Did God make doors, toys, and cars?
- Did God make hats so the Sun does not beat down on us?
- Did God make houses?
- Did God make the balls that we play with?
- Did God make TVs, and what TV shows does He watch?
- Do all animals obey God?
- How and why was darkness born?
- How are clouds made?
- How did plants sprout from the Earth for the first time, and why are they mostly green?
- How were human beings made, and where do babies come from?
- How did the first human beings look after they were formed?
- How were bones made?
- Why do birds fly, but people cannot?
- How was the air we breathe made?
- How was the sky made, and why is it blue and black sometimes?
- How was time born, and why are the morning, the afternoon, and the evening different?
- How were angels formed?
- How were animals formed?
- How were seeds made?
- Why are most plants green?
- How were the seas and the oceans made?

What if Dad can help you get closer to answering all of those questions?" Wouldn't it be cool if you could get the accurate answers to all of them? This book, "*How Baby Universe Was Born*," is not the only book that my dad wrote. As I told you earlier, he wrote another children's book called "*How God Created Baby Universe*," in which he addresses all those questions about God that those who believe in Him or want to know more about Him will appreciate. Get that book today and read it.

Maybe you, or your parents, or the adults in your house are wondering:

- Does the Bible scientifically teach anything else about the universe's origin that most people ignore?
- Is there any way for Christians to talk to evolutionists, Big Bang proponents, atheists, and all other freethinkers about the formation of the universe, and they will be very interested in knowing more about God?

Science180: All the Universe-Origin and Life-Origin Solutions You Love

- Is there a best way for believers to talk to stubborn rationalists about the formation of the universe so they beg to be led to God right away?
- Do you have to embrace evolution or deny God to scientifically prove that God created the universe in 6 literal days?
- Do Christians have to compromise with evolutionism to convince anyone about creation or the existence of God?
- Do you have to stop thinking to scientifically prove that God created the universe?
- Can anyone fix the trend according to which more people are denying God at the expense of secular theories because they think that it is impossible for science and faith to meet?
- Can anything be done to scientifically fix what is causing more and more college students to question the Bible, abandon their Christian faith, while kissing secular biology and physics books that program them to believe in evolutionism, Big Bang, and other theories that deny God?
- Is a church or a pastor making you doubt God or the Biblical story of creation?
- Is your school teaching about the origin of the universe, making you doubt creation?
- Is science making you doubt your faith?
- Is the Biblical account of creation making you doubt science?
- Is science making you doubt God or the Bible?
- Are science and the Bible really diametrically opposed or in conflict with each other?
- Is it true that the science versus the Bible debate will never be settled?
- Are science and the Bible really incompatible?
- Does the Bible really disagree with science?
- Is the Bible actually an obstacle to scientific progress?
- Is science being used to deny God?
- Can science affect our mind and faith?
- Why do secular rationalists and freethinkers think that Christians are irrational?
- What if Christians are not as irrational as secular rationalists and freethinkers think they are?
- What process can rationally explain the Biblical account of creation and remove all roadblocks in the way of previous scientists who tried to scientifically interpret the Genesis story?
- Is there any lie that most pastors spew from the pulpit about creation, and is there any simple scientific formula to accurately overcome it quickly without angering Christians, God, and the unbelievers?

Nathanael-Israel Israel: Author of "Reconciling Science and Creation Accurately"

SECTION 2: HOW THE GALAXIES, THE PLANETS, THE MOON, AND THE SUN WERE FORMED

- Can Biblical creation fuel your scientific success?
- Can creationists disagree with anti-creationists without angering anyone?
- Can atheists, rationalists, secularists, and freethinkers scientifically survive and heal from doubting God?
- What is the best way to think fully within the conflict between reason and faith?
- Which Biblical verses have the power to bring science to its knees—or can't they?
- Why do most people believe in and stick with incorrect universe-origin theories?
- Why does the discovery of the perfect link between science and Biblical creation of the universe mean good news for atheists, Evolutionists, and Big Bang proponents?

Besides the children's books, two other books that my dad wrote about the universe creation that can help you answer many of these questions are:

- *"Reconciling Science and Creation Accurately"*
- *"From Science to Bible's Conclusions"*

Later, I will talk about them, how people can get my dad to answer their questions, and more. See also www.Science180.com/speaking.

Nathanael-Israel Israel: Author of "Reconciling Science and Creation Accurately"

SECTION 4: RESOURCES FOR YOUR PARENTS OR THE ADULTS IN YOUR HOME

24. OTHER AMAZING BOOKS WRITTEN BY MY DAD, NATHANAEL-ISRAEL ISRAEL, THAT CAN BENEFIT YOU

'Science180 Academy' Success Strategy
SCIENCE180 PUBLISHING: AUTHORS WANTED

Science180 Publishing, the American publishing company that published the groundbreaking discovery about the origin of the universe, of life, and of chemicals spearheaded by Dr. Nathanael-Israel Israel, really wants to publish your book(s) regardless of your field of expertise. This is a unique opportunity for:

- established authors
- people aspiring to become authors
- people who have written a book or are wanting to write one and need help with anything regarding publishing
- people who are not well known, inexperienced
- people whose books are viewed as nonconformist, controversial, or unconventional
- people who do not have enough resources or knowledge to navigate the publishing process
- people who are struggling to find an affordable, experienced, and high-quality publisher

Although Science180 Publishing is based in the USA, it can publish your books within your budget regardless of your geographical location. Science180 Publishing is highly interested in your document and possibly helping you publish it. Please visit Science180Publishing.com to explore how we may assist you. No matter the content of your book, as far as it is original, not promoting anything illegal, not duplicating anyone else's idea, Science180 Publishing can help you publish it in the USA. Please contact us asap and see how we can help. To start your journey of publishing your book with Science180 Publishing, please visit Science180Publishing.com today.

'Science180 Academy' Success Strategy:
SCIENCE180 BOOKS THAT WILL HELP YOU OR YOUR PARENTS!

I, Nathanael-Israel Israel, broke down my discovery about the formation of the universe into many books so that you, the readers, can pick the ones that correspond to your needs and interests without disappointing you or wasting your precious time. These books come in many versions (e.g., scientific version, public version, chemical version, biological version, biblical or prophetic version, pseudepigraphic version, and a children's version) targeting people according to their expertise, educational background, and interests as briefed below:

1. *"TURBULENT ORIGIN OF THE UNIVERSE"* (This is the scientific version of my book tailored to scientists and anyone interested in the detailed scientific demonstration of the universe formation). In this book I used the "mother of all turbulences" to scientifically demonstrate the formation of the universe so that scientists can understand and reorient the course of their research, teaching, and publishing and accept the truth to better live today and forever. Get *"Turbulent Origin of the Universe"* today to begin an incredible journey of accurately decoding the universe and change your life forever! Learn more at Science180.com/scientific

2. *"RECONCILING SCIENCE AND CREATION ACCURATELY"* (this is the book that I called the "Biblical or prophetic version of my book on the universe's origin, and it targets Christians and anyone interested in knowing the Biblical perspective of the creation of the universe). This important book accurately demonstrates the marvelous creation and formation of the universe by God in six consecutive 24-hour-days, and answers many questions about the universe's creation so that after acknowledging Him (who deserves all the glory now and forever), human beings can choose life and avoid the terrible judgment awaiting the unbelievers in the world to come. Get this thoughtful book now to figure out what happened at the beginning, what is coming up, and why it is time to urgently rethink everything you have been told about the universe's origin so you don't eventually regret! Don't say I did not tell you! Learn more at Science180.com/biblical

3. **"*TURBULENT ORIGIN OF CHEMICAL PARTICLES*"** (Called the "chemical version" of my book on the universe's origin, this elegant book targets chemists, biochemists, and anyone interested in chemistry). With *"Turbulent Origin of Chemical Particles"*, the accurate decrypting and understanding of the formation of chemicals has never been profitable and easy. Hence this great book is THE ultimate how-to guide for great people wanting to correctly decode the origin of the chemicals and positively transform their lives. Get this celebrated book today. Learn more at Science180.com/chemical

4. **"*ORIGIN OF THE SPIRITUAL WORLD*"** (This book is what I called the pseudepigraphic or hidden version of my books on the universe origin, and it is meant for believers who want to tap into a higher level of scriptural secrets that most people may not believe). This book draws the attention of the world toward the pseudepigrapha (a collection of hidden and rejected books, yet filled with deep secrets still valuable today) and explaining how, for thousands of years, God has already revealed deep details about the supernatural origin of the universe, but people (including those who believe or claim to believe in Him) have just refused to literally accept God's mysterious story of creation, which can never be understood by just sticking with conventional science. If you believe in God but have some origin-related questions whose answers you cannot find anywhere, not even in the Bible, and if you want to tap into historically neglected revelations to answer fundamental universe and life questions, then be sure to get a copy of *"Origin of the Spiritual World"* today. Learn more at Science180.com/pseudepigrapha

5. **"*FROM SCIENCE TO BIBLE'S CONCLUSIONS*"** (I called this book the "public version" of my book on the origin of the universe and it is tailored for the general public, and it is a great summary of the scientific version from a perspective that laypeople will fully understand). In this book, I, Nathanael-Israel Israel, broke down the complicated (scientific, philosophical including religious) data about the origin of the universe in a simple language that the general public can fully understand, and know in order to live happily forever. Quickly grab and read this scientifically verifiable, bestselling book to finally get the accurate, jaw-dropping answer that has been rationally shaking believers, skeptics, and all freethinkers. Don't wait! Learn more at Science180.com/public

6. **"TURBULENT ORIGIN OF LIFE"** (This is the biological or life version of my book on the origin of the universe). It is meant to suit scientists, nonscientists, and all kids of laypeople, and it decodes the origin of all forms of life so human beings can understand and better live. As of 2025, this book is my only book devoted to the origin of all forms of life, and it will help you to grasp in a simple language what is needed to fully understand the formation of all forms of life. Whether you are a scientist or a layperson, a believer, or a skeptic, you cannot afford to ignore the greater, better, faster, simpler, cheaper, easier, and accurate formula unlocked in this important book that successfully decoded the origin of life. Get *"Turbulence Origin of Life"* today and change lives. Don't wait. Learn more at Science180.com/life

7. **"HOW BABY UNIVERSE WAS BORN"** (How was the universe formed? Did God really form it like some people believe, or did it come out of some long processes? How can we scientifically prove and break down this difficult mystery in a language that children will fully understand and like?) Get the answers as you read this book that I called the "children version" of my book on the origin of the universe and life. Accurately explaining the complex formation of the universe and of life to children can be very hard in our modern world, but by getting *"How Baby Universe was Born"*, you will know the proven formula to help children to easily understand their huge universe-origin and life-origin questions with confidence, humor, and joy. They will surely belly laugh and thank you for it! It is time to buy this pragmatic book and offer it to the children in your life today. Learn more at Science180.com/children

8. **"HOW GOD CREATED BABY UNIVERSE"**. The most difficult part of writing scientific things to children is how to break down complex technical concepts into simple words that they and even anyone who can read and clearly understand (without losing the accurate details and facts). When the topic to address is about the origin of the universe, the task is even more challenging for most people, but not for Nathanael-Israel Israel. As long as you can read, you will find this amazing book extremely helpful to grasp all complicated concepts needed to properly crack the origin of the universe in a language that even children ages 7-12 and anyone who did not go very far in school can fully comprehend.

9. *"SCIENCE180 ACCURATE SCIENTIFIC PROOF OF GOD"* (Whether you are a believer, an unbeliever, a freethinker, an administrator, a politician, a curriculum designer, a curriculum specialist, an education policymaker, a librarian, a school board member, a parent, a researcher, a student, a teacher, clergy, or a layperson, as long as you are really seeking to scientifically understand the rational proof of the existence of God, *"Science180 Accurate Scientific Proof of God"* is the much-admired book written for great people just like you). As long as you are interested in the first and the only scientific book that talks to anti-creationists, evolutionists, big bang proponents, atheists, and all other freethinkers and rationalists about the universe's formation and they bigly beg to know more about God, the creator, whom they mistakenly deny; then this book is for you. As long as you are really seeking to scientifically understand the rational proof of the existence of God, *"Science180 Accurate Scientific Proof of God"* is the much-admired book written for great people just like you. Grab it today and start reading it. Don't wait any longer! Learn more at Science180.com/godproof

If you want to have the entire big picture of my discovery of the origin of the universe, life, and chemicals, and to enlighten your life and career, then plan to get all or some of these books that best suit your needs and interests. For more details, visit Science180.com/books

Below are more details on each of these books.

Another Book by Nathanael-Israel Israel:
HOW GOD CREATED BABY UNIVERSE

THE FIRST AND ONLY BOOK THAT ACCURATELY EXPLAINS EVERYTHING ABOUT THE FORMATION OF THE UNIVERSE AND LIFE IN A WONDERFUL LANGUAGE THAT ALL CHILDREN AGES 7-12 CAN EASILY, FULLY UNDERSTAND & ENJOY!

As the only universe-origin book that your whole family will like and enjoy together, *"How God Created Baby Universe"* will set children on the path of success by accurately showing them early in life the formation of the universe, and how to detect errors in theories or stories that would misguide them as they grow up. Therefore, you need to add this great, efficient, trustworthy, and cost-effective book to the strategic journey of children toward their best tomorrow. With *"How God Created Baby Universe"*, you will:

- Have a peace of mind that children will get accurate, fit, and easy to understand universe-origin information that will produce real results in their life

- Become the leader that captures the heart of children craving for the original explanation of the formation of the universe so you can clear their way for freedom, power, technology, innovation, and breakthroughs of the future (learn more at Science180.com/children)

- Protect yourself and loved ones from wrong theories in the literature and the media by keeping children secured and empowered with the true knowledge of how the universe began

- Explain complicated secrets to children about how to locate mistakes in origin-related theories so you can save time, money, and other resources to improve their lives

- Ultimately boost children's confidence in detecting, confronting, and avoiding wrong theories by knowing the facts and real processes involved in the formation of the universe

- Help children to easily sort out their origin-related questions using strategies that get them to tap into deep secrets that even highly educated people ignore

- Clearly explain to children how to mathematically know without a doubt whether God created the universe as the Bible says or billions of years evolution processes formed it

Accurately explaining the complex formation of the universe and of life to children can be very hard in our modern world, but by getting *"How God Created Baby Universe"*, you will know the proven formula to help children to easily understand their huge universe-origin and life-origin questions with confidence, humor, and joy.

They will surely laugh aloud while reading this book and thank you for it! It is time to buy this pragmatic book to help the children in your life today.

Member of the American Association for the Advancement of Science, American Chemical Society, and the American Society for Microbiology, **Dr. Nathanael-Israel Israel is** a Beninese-American scientist and international consultant, who shows the world how to scientifically decode the formation of the universe, of life, and who is known as the creator of the Chemicals Turbulent Origin Formula™, the inventor of the Life Turbulent Origin Formula™, and the discoverer of the Universe Creation Formula™. Learn more at Israel120.com.

Another Book by Nathanael-Israel Israel:
TURBULENT ORIGIN OF THE UNIVERSE

THE FIRST AND ONLY SCIENTIFIC BOOK THAT ACCURATELY EXPLAINS EVERYTHING YOU NEED TO UNCONVENTIONALLY, EASILY, AFFORDABLY, AND ENJOYABLY DECODE THE UNIVERSE FORMATION

In *"Turbulent Origin of the Universe"*, filled with great diagrams and digestible scientific facts, you will discover, learn, or get:

- The all-in-one, proven & uncomplicated scientific formula that accurately decoded the formation of the universe, and that explained the birthdate of the stars, planets, satellites, asteroids, and all other celestial bodies in the universe, so you can position yourself to stay on top of your competitors and avoid repeating crucial mistakes that many people have ignorantly made at their own perils

- Extraordinary, unprecedented, accurate insights into the first factors (e.g. early universe physics) that defined the history and formation of the universe so you can tap into deep scientific secrets you ignore, and set yourself apart from others

- The new physics that will revolutionize science forever and land you into a zone of original ideas that improve lives nonstop regardless of your expertise

- The 4 simple things without which it is impossible for anyone to ever understand the formation of the universe, think accurately, work differently, achieve, or perform better for superior results

- The verified key to move the cosmological mountains of misunderstanding, so you can confidently free your mind from doubts, improve your health, and prevent you from any danger connected with sticking with wrong assumptions

- Save time and money, and enjoy your life once you remove errors holding your true understanding of the universe's origin captive

- Historic scientific proof of whether a planet was formed in 2.82 days, whether a satellite was formed in 3.32 days, and whether a star was formed in 3.69 days after the beginning of the universe; so you can creatively produce and address a broader work spectrum by learning how to effectively communicate with and establish unusual connections between otherwise disconnected and disparate scientific data

Science180: Bringing People Together Through the Power of the Accurate Decoding and Understanding of the Universe Creation

- The scientific formula that successfully tested the existence of God in a way that shocked believers, skeptics, and all other freethinkers
- Why the scientific community has failed to sufficiently explain the origin of the universe and understand how existing theories have missed and undefined central ideas, and imposed limits on the vision of scientists
- Specific in-depth knowledge, up-to-the-minute information, and ideas so you can expand your market, cut useless costs, stop wasting time on inadequate projects, and start focusing on the profitable solutions (Science180.com/scientific)
- How Science180 Academy can strategically enlighten you and guide you to navigate and filter the massive data collected on the universe, so you can answer the world's most challenging questions, remove any scientific and philosophical cataracts that may be blocking you, and bring you many steps closer to your best life
- How to better resonate with your target market that is craving something original that breaks wrong explanations of the universe's origin

Get *"Turbulent Origin of the Universe"* today to begin an incredible journey of accurately decoding the universe and change your life forever!

Dr. Nathanael-Israel Israel is told by people that he is the #1 Universe-origin, Life-origin, and Chemicals-origin expert. He is the founder of Science180 and the author of many books on the origin of the universe and its content. To learn more about how he may help you, visit Israel120.com.

Another Book by Nathanael-Israel Israel:
RECONCILING SCIENCE AND CREATION ACCURATELY

THERE IS ONLY ONE SIMPLE, COMPELLING, SOLUTION-DIRECTED SCIENTIFIC FORMULA ACCURATE ENOUGH TO RATIONALLY EXPLAIN HOW GOD CREATED THE UNIVERSE

"Reconciling Science and Creation Accurately" is a landmark book in universe-origin writing from a rare perspective by one of the most respected minds of our time. It scientifically explores the most challenging questions of all times that believers, nonbelievers, and all freethinkers are interested in: How can we rationally demonstrate, without checking our brain at the door in the name of faith, that God created the universe? How did the universe begin and what processes did God use to create it? Are these processes still operating in the universe or not? Can believers abandon wrong theories if they think it is impossible for science to literally prove the Genesis story, or if they think that science is evil and diametrically opposed to faith, or if they compromisingly embrace scientific theories that contradict the Biblical account of creation written before the scientific era? What can believers do to help the skeptics believe in the Biblical narrative of creation?

Lucky you, Dr. Nathanael-Israel Israel successfully navigated all those questions with an accuracy that both scientists and nonscientists have been applauding across the globe. After reading *"Reconciling Science and Creation Accurately"*, you will confidently:

- Scientifically prove the Biblical account of the creation of the universe and the existence of God in a way that makes the head of those who deny God to spin faster than a DJ's turntable
- Know how to rationally talk to anti-creationists, evolutionists, Big Bang proponents, atheists, skeptics, and other freethinkers about the universe's formation and they will beg you to know more about God, the Creator, that they mistakenly rejected
- Discover very accurate, rare, and factual truths about the universe's origin that will save you time and money, and get you much closer to the better and joyful life you want to live today and forever
- Improve your health and faith by knowing that the existence of God can be scientifically justified using Science180 Cosmology and particularly Science180 Creationism
- Enter a new area of freedom and power by crushing the head of and breaking free from the suffocating expectations of all wrong theories that have hijacked secular and religious education, and that have held the Biblical account of creation captive for almost 3500 years

Science180: Bringing People Together Through the Power of the Accurate Decoding and Understanding of the Universe Creation

- Break free from the suffocating expectations of some forms of creationism that have sequestered the mind of some believers for a long time
- Uncompromisingly, intelligently, and scientifically explode the myth of those who, instead of literally taking the Biblical days of creation as 24-hour consecutive days, think that they were millions of years, or were representative of long ages, or that millions of years existed before them or were positioned between them
- Understand the accurate standard to interpret the Biblical account of creation thanks to Science180's breakthrough that transformed science and laid a foundational bedrock for the inerrancy of Scripture

Now that Genesis (the oldest manuscript in the world, written before science and most religions were born) is scientifically proven to be correct (Science180.com/biblical), what unstoppable, jaw-dropping paradigm shift will the discovery of the perfect alignment between science and the Bible bring into the religious, rational, and secular world today? Get this thoughtful book now to figure out what happened at the beginning, what is coming up, and why it is time to urgently rethink everything you have been told about the universe's origin so you don't eventually regret! Don't say nobody told you!

Founder of Science180 Academy, **Dr. Nathanael-Israel Israel** is acknowledged worldwide as the discoverer of the all-in-one, proven, and simple scientific formula that accurately cracked the origin of the universe, of life, and of chemicals and that scientifically unearthed the holy grail at the intersection of science and the Biblical account of creation. Learn more at Israel120.com.

Another Book by Nathanael-Israel Israel:
TURBULENT ORIGIN OF CHEMICAL PARTICLES

FIND ALL THE RELIABLE, CONVINCING, SCIENTIFIC ANSWERS YOU NEED TO SUCCESSFULLY DECODE THE ORIGIN OF CHEMICAL PARTICLES SAFELY

Where did all elementary particles and composite particles including atoms, molecules, minerals, and rocks, come from? What are the fundamental factors, the machinery, and the generic processes that defined their formation and properties? What was the nature of their precursors at the beginning of the universe and what underlying processes shaped or molded them into the chemicals we know today? What was the primary cause of the abundance and diversity of chemicals in the celestial bodies in the universe? What is the accurate link between the formation of chemical particles and the formation of galaxies, stars, planets, asteroids, and satellites? What light can the origin of chemicals shed on the real cause and meaning of gravity and the other so-called fundamental forces in nature? How does the formation of the chemical particles fit into the big picture of the formation of the universe?

After studying these questions for more than 12 years, Dr. Nathanael-Israel Israel discovered that the proper understanding of the origin of chemical particles is a very challenging but profitable task that requires original, scientific, mathematical, and philosophic efforts beyond the current state of modern science—until recently. The solution for all of these puzzling problems: *"Turbulent Origin of Chemical Particles"*, the straightforward and trustworthy book that will help you to quickly, cheaply, easily, and efficiently navigate everything you need to know to finally solve the hard problems about the origin, the formation, and the functioning of all chemical particles. Whether you are a chemist, a biochemist, any other scientist, or an engineer, as long as you have a reasonable background in chemistry but ignore how to scientifically demonstrate the origin of all chemical particles, this marvelous book is for you! Amazingly packed with eye-popping analysis, fantastic graphs, tables, and the historic formula that broke the universe-origin code, *"Turbulent Origin of Chemical Particles"* will:

- Make it easier than ever for you to properly understand, decrypt, and articulate the real origin of natural chemical particles in the universe, therefore freeing you from false and boring explanations of the origin of all matters, and embrace the proven theory that opens doors to unparallel opportunities
- Professionally teach you how to transform the true knowledge of the origin of chemical particles into insights that significantly add value to your life in less time, and successfully establish you as a symbol of freedom, power, creativity, and originality in your field of expertise

Science180: Bringing People Together Through the Power of the Accurate Decoding and Understanding of the Universe Creation

- Fire you up to become the best version of you, and to cause positive changes to your initiatives that will profit you nonstop
- Discover thrilling illustrations and unconventional explanations of the formation of all matter in the universe, written in a simple language that brings humankind much closer to the complete deciphering of the mysteries at the very heart of chemistry, and open the way to a future of technology, innovation, discoveries, and breakthroughs
- Equip you to bypass technical knowledge that restricts non-experts from accessing the origin-related secrets contained in the massive scientific data, and get to the bottom of origin-related mysteries regardless of your background so you can empower yourself to leave unforgettable marks in your field of expertise
- Learn more at Science180.com/chemical

With *"Turbulent Origin of Chemical Particles,"* the accurate decrypting and understanding of the formation of chemicals has never been profitable and easy. Hence this great book is THE ultimate how-to guide for great people wanting to correctly decode the origin of the chemicals and positively transform their lives. Get this celebrated book today. Don't wait!

Known as the nonconformist, rule-breaker, and accurate demonstrator of the universe's origin, **Dr. Nathanael-Israel Israel** is the founder of Science180, the one-stop for answering the most crucial universe and life's origin questions. He has had the honor to be acknowledged as the fearless universe-origin decryption trailblazer. Learn more at Israel120.com.

Another Book by Nathanael-Israel Israel:
ORIGIN OF THE SPIRITUAL WORLD

ONLY ONE ANCIENT BLUEPRINT HAS THE RELIABLE POWER TO HELP YOU TO ACCURATELY DECRYPT THE SPIRITUAL ORIGIN AND HISTORY OF EVERYTHING IN THE UNIVERSE

Countless books talk about the origin of the universe and of life, but this amazing book is the first and the only one that has undeniably explained how the formation of the universe and everything in it was truly revealed in the rejected and hidden scriptures such as the Books of Enoch and others. In *"Origin of the Spiritual World,"* you will:

- Discover deep-rejected secrets that have prevented humankind from unearthing the beginning of the universe
- Plainly see the scientific proof (hidden in scriptures) of the formation of the Earth, the Moon, and the Sun in a matter of days, a historic revelation that bizarrely and shockingly matches the scientific data as scientifically proved in *"From Science to Bible's Conclusions"*, a popular book written by Dr. Nathanael-Israel Israel
- Properly use the lost and rejected scriptures to articulate the process by which the universe was formed, and use that insight to improve your understanding of the Bible, innovate in your domain of interest, and improve your life perpetually
- Empower and align yourself with the historic breakthrough that has done what no other discovery has ever done: accurately unlock and decode mysteries concerning the origin of the cosmos and its content using scientific keys revealed in ancient scriptures that some elites have concealed (Science180.com/pseudepigraphic)
- Discover and apprehend the complex formation of the universe and life without leaving out the challenging questions that people of all ages have been struggling to answer for thousands of years, while the answers were hidden
- Find more joy in life through a clear interpretation of old and fresh revelations about the creation of the universe astonishingly backed by modern science, which some people wrongly think opposes the Bible
- Make a difference and blaze new trails for those who depend on your leadership

If you believe in God, have some origin-related questions whose answers you cannot find anywhere, not even in the Bible, and if you want to tap into historically neglected revelations to answer fundamental universe and life questions, then be sure to get a copy of *"Origin of the Spiritual World"* today.

Dr. Nathanael-Israel Israel happens to be the discoverer of the historic mathematical equations that scientifically demonstrated that the Earth was formed 2.82 days, the Moon 3.32 days, and the Sun 3.69 days after the beginning of the universe, therefore confirming the Biblical account of creation that revealed about 3500 years ago that the formation of the Earth was completed on the 3rd day, while that of the Moon and the Sun was completed on the 4th day of creation. Nathanael-Israel Israel is referred to as the "Undisputable Specialist of all Questions at the Intersection of Science and Biblical Creation". Learn more about this rare scientist at Israel120.com.

Another Book by Nathanael-Israel Israel:
FROM SCIENCE TO BIBLE'S CONCLUSIONS

THE # 1 UNIVERSE-ORIGIN MASTERPIECE OF ALL TIME … AND THE MOST ACCURATE SCIENTIFIC FORMULA THAT STOOD AND WILL STAND THE TEST OF TIME AND OF MATHEMATICS

The real reason scientists have been struggling to accurately understand the universe-formation is because they have spent centuries collecting expensive, complicated, and massive amounts of data, but learned very little, if not nothing, about how to unconventionally step back to properly analyze it to decode the universe. Consequently, people learned to collect all kinds of data everywhere to build models and imaginary concepts that betray their discernment, but they never learned to unlearn wrong theories, nor learned how to stop trashing great raw data hidden in theories they dislike or misunderstand; they never knew where to find and how to properly combine the fundamental variables without which it is impossible to ever clear the way so their data can properly work for and precisely lead them to the real origin of the universe. How can people abandon the dangerous theories they think are correct because they don't know any better ones?

Lucky you, that is where Dr. Nathanael-Israel Israel, the founder of Science180 (Science180.com) came in to properly reanalyze and put under control these costly, underrated data to provide the accurate and simple solution people have been looking for throughout the ages, but that they have ignored.

In *"From Science to Bible's Conclusions"*, you will:

- Get a world class explanation of the 4 fundamental variables without which it is unquestionably impossible to ever decode the universe-formation scientifically

- Save time and money, and enjoy a life filled with the wonderful peace that the accurate understanding of the universe's origin can create

- Discover the errors in the scientific theories and religious belief systems about the universe-formation that are putting you at risk, and learn how to take control over cosmological threats lurking at the edge of your rational mind, faith, disbelief, or doubt

- Unlock the accurate scientific formula to rationally test the existence of God in a historic way that uncompromisingly satisfies both believers and skeptics (Science180.com/public)

- Get all you need to become a knowledgeable person who will never again need anybody else to explain to you the origin of the universe, for, you will fully understand and articulate it yourself and rationally know whether science is really at war with religion

- Receive deep insights that even those who went to university for years were not able to decrypt by themselves, so you can equip yourself to eliminate all forms of scientific and religious universe-origin prejudices
- Discover whether the scientific data finally confirms that the formation of the Earth was completed on the 3rd day, while that of the Moon and the Sun was on the 4th day of creation like the Bible says, or whether the data proves that it took billions of years to progressively form the universe
- Understand the celebrated scientific formula that rationally puts to rest all debates about the relationship between science, faith, and all theories about the universe's origin so you can properly develop yourself, expand your network, and shape your future

Quickly grab and read this scientifically verifiable, bestselling book to finally get the accurate, jaw-dropping answer that has been rationally shaking believers, skeptics, and all freethinkers. Don't wait!

Dr. Nathanael-Israel Israel has had the honor to be acknowledged as the #1 universe-origin, life-origin, and chemicals-origin expert. He is the author of *"Turbulent Origin of the Universe"*, *"Reconciling Science and Creation Accurately"*, *"Turbulent Origin of Chemical Particles"*, *"Turbulent Origin of Life"*, *"How Baby Universe Was Born"*, *"Science180 Accurate Scientific Proof of God"*. Visit Israel120.com to learn more about this world's most trusted expert that helps scientists and laypeople to properly decode the origin and formation of the universe, life, and chemicals so people can live more effectively nonstop.

Another Book by Nathanael-Israel Israel:
TURBULENT ORIGIN OF LIFE

THE ONLY ACCURATE FORMULA TO SCIENTIFICALLY EXPLAIN THE FORMATION OF ALL FORMS OF LIFE QUICKLY

Every human being will benefit from understanding the real origin of life. But the problem is that most efforts to explain the origin of life are complex, inaccurate, confusing, partisan, and complicated, therefore, creating serious challenges to those who are eager to scientifically decrypt where all forms of life came from. Most people want an accurate, simple, straightforward, nonpartisan life-origin book that is free from jargon and difficult concepts only known by the experts. This elegant scientific book breaks down the technicality of the origin of life in a language that even the nonscientists can easily comprehend. It is a trustworthy book that will help you to quickly, cheaply, easily, and efficiently navigate everything you need to know to finally decode and solve the puzzling problems about the origin of life, while also giving you a crash course on the universe's origin.

Unlike any book you have ever read on the origin of life, this historic masterpiece (that distills complex scientific data down to simple explanations that make sense) is the starting point of any smart person wanting to rationally understand the formation of all living things. By the time you finish reading "*Turbulent Origin of Life*", you will discover:

- Why in spite of the massive amount of scientific data collected on living things, scientists have misunderstood the formation of life until now, and then uncover in a simple language the one thing that was needed to accurately crack the code of life but that scientists have missed and that has been causing them headaches, overwhelm, and burnout
- Step-by-step pathway to decode the origin of life and get the power, freedom, and boldness to take advantage of the opportunities that accurate understanding of the origin of life creates (Science180.com/life)
- The high connection between the code of the universe formation and the process by which life on Earth was formed so you can become a fulfilled thought leader in your field of expertise
- Tools to stand as a lighting bolt that electrifies those who are still struggling to understand the formation of all forms of life in the universe
- Strategies to push the boundaries of human abilities to properly understand what is perceived as un-understandable, mysterious, supernatural, unimaginable, impossible, and unthinkable that hold people back

- Scientific approach to holistically detect, correct, and remove all misinformation, ambiguity, and misleading claims and theories surrounding the origin of life

Whether you are a scientist or a layperson, a believer or a skeptic, you cannot afford to ignore the greater, better, faster, simpler, cheaper, easier, and accurate formula unlocked in this important book that successfully decoded the origin of life. Get *"Turbulence Origin of Life"* today and change lives! Don't wait!

Dr. Nathanael-Israel Israel is the Father of Science180 Cosmology and the Founder of Science180 Academy. He is fortunate to be known as the source of unconventional wisdom and knowledge that help people accurately crack the code of the formation of the universe, of life, and of chemicals. Get some resources by visiting his personal website at Israel120.com.

Another Book by Nathanael-Israel Israel:
SCIENCE180 ACCURATE SCIENTIFIC PROOF OF GOD

THE FIRST AND THE ONLY SCIENTIFIC BOOK THAT TALKS TO ANTI-CREATIONISTS, EVOLUTIONISTS, BIG BANG PROPONENTS, ATHEISTS, AND ALL OTHER FREETHINKERS AND RATIONALISTS ABOUT THE UNIVERSE'S FORMATION, AND THEY BEG TO KNOW MORE ABOUT GOD, THE CREATOR, WHOM THEY DENY.

As you read this historic book, you will:

- Scientifically know what is the one clear sign you should always pay attention to in your efforts to decipher the primary cause and the key drivers of the fundamental processes responsible for universe formation.
- Discover the only way to scientifically know if God exists and, if so, which of the thousands of beings worshipped across the globe is the true God
- Accurately answer the most critical universe-origin and life-origin questions so you can stop standing in tension with consequential question marks, including those related to religion and reason or the so-called war between science and the Bible
- Discover the errors in the scientific and religious theories about the universe-origin and life-origin that are putting you at a high risk you will never recover from if you don't quickly and confidently learn how to rationally take control over threats lurking at the edge of your efforts to understand the universe and life today
- Challenge the cosmological status quo and embrace the real change that will disrupt the hidden cages that may be holding you and that you ignore
- Definitively answer all your doubts about the source or author of the universe and life … (learn more at Science180.com/godproof)
- Understand that religion or faith, reason or science can coexist and can be properly reconciled to accurately lead you to the correct source of everything in the universe
- Satisfy your burning desire for freedom from beliefs and scientific theories about the universe's origin and life-origin that suffocate you and bind your mind, faith, unbelief, heart, and education
- Scientifically set on fire all false theories or dogmas about the existence of God, the Creator, that are enslaving humankind

Science180: Bringing People Together Through the Power of the Accurate Decoding and Understanding of the Universe Creation

Whether you are a believer, unbeliever, freethinker, administrator, politician, curriculum designer, curriculum specialist, education policymaker, teacher, librarian, school board member, researcher, parent, student, clergy, or a layperson, as long as you are really seeking to scientifically understand the rational proof of the existence of God, *"Science180 Accurate Scientific Proof of God"* is the much-admired book written for great people just like you! Grab your copy today and start reading it! Don't wait any longer!

Dr. Nathanael-Israel Israel is a Beninese-American scientist, entrepreneur, and international consultant, who shows people of all ages and educational backgrounds how to scientifically decode the formation of the universe and of life, and who is acknowledged as the creator of the Chemicals Turbulent Origin Formula™, the inventor of the Life Turbulent Origin Formula™, and the discoverer of the Universe Creation Formula™. He is the Founder of Science180 Academy, which is trailblazing the reconciliation between science and the creation.

25. RESOURCES ABOUT SCIENCE180 FOR YOUR PARENTS OR THE ADULTS IN YOUR HOME

'Science180 Academy' Success Strategy:
SCIENCE180 INTERVIEW REPORT (AKA SCIENCE180 INTERNET-TV-RADIO INTERVIEW REPORT)

Science180 Interview Report is the newsletter to read for guests and unconventional show ideas at the intersection of science and faith. Indeed, many hot questions are still unanswered on the road leading to the correct understanding of the origin of the universe, of life, and of chemicals. But most people don't know where to find the accurate answers to those challenging questions. What if, with one simple call, you can accurately answer all of those questions? You need to get in touch with or interview Dr. Nathanael-Israel Israel on your show, radio, tv, podcast, and even website, or invite him for a live presentation at your organization if your audience can benefit from any of the following show, talk, speaking, or interview ideas:

- Can a single variable play a crucial role in cracking the universe?
- Can the distance separating celestial bodies give a clue to the universe's birthdate?
- What is the master key to crack the universe-origin?
- What is the little-known variable that hides a key to understand something unique about the movement of celestial bodies?
- How to deal with the fear of not knowing the origin of the universe and life?
- Can you be really free from doubt about the universe-origin?
- Can you be really free from doubt about God's existence?

- Can mathematics and science collide to accurately explain the creation of the universe, of life, and of chemical particles?
- Can mathematics help science to crack the code of the origin of life?
- Can we mathematically prove that the formation of the Earth was completed on the 3rd day of creation like the Bible says?
- Can we scientifically demonstrate without a doubt that the Moon and the Sun were really formed on the 4th day of creation like the Bible says?
- Can mathematics and science rescue Christians in their efforts to rationally prove the existence of God, the Creator?
- Did the Quran and any other religious book make any gigantic error about the universe creation that any scientific formula proves the Bible got right—and vice versa?
- Why the secularist world doesn't care much if Christians and their leaders believe in Evolutionism, but they actually care much if they don't believe in the billions of years process?
- Is the Bible's account of creation making you doubt science?
- Does the Bible scientifically teach anything about the universe-origin that most people including Christians ignore?
- Must Christians apologize to atheists, rationalists, and all freethinkers for the proofs creationist scholars and preachers have used to demonstrate creation?
- Can anyone really be scientifically 100% sure and prove that God created the universe?
- Can we explain the formation of the universe through natural processes without evoking evolution and Big Bang?
- Is it a waste of time to attempt to prove the Biblical creation using science or historical investigation?
- Is there any need to prove the Biblical creation to be true?

I know you may be tempted to answer these questions by yourselves, but avoid landing yourself on wrong paths that caused some people to lose contact with reality, it is better to get the accurate answer from the know-how expert, Dr. Nathanael-Israel Israel, the author of many books on the origin of the universe, of life, and of chemicals, and the standout expert who accurately decoded the scientific formula that forces science to bow to the truth. If you would like to register to Science180 Interview Report so we can periodically send you show ideas and opportunities related to the origin of the universe, of life, and of chemicals particles, please visit Science180InterviewReport.com for more details. Contact Nathanael-Israel Israel (visit www.Israel120.com) if you want to get his answers for any of these questions.

'Science180 Academy' Success Strategy:
SCIENCE180 ACADEMY

Science180 Academy is a training, speaking, consulting, and mentoring program designed to groom and empower people of all backgrounds in the truth about the origin of the universe, life, and chemicals. According to their background and interest, trainees are taught different levels of scientific facts to grasp a deeper understanding of the origin of the universe, how to properly think to unearth mysteries hidden in the massive scientific data collected across the globe but which is unfortunately less analyzed. If you want to be enlightened and equipped so you can cause positive changes in your respective field of expertise, then Science180 Academy program is for you.

Science180 Academy does not confer college credit, grant degrees, or grade its attendants, participants, or students. It is not an accredited university or college, but is the one-stop-destination for universe-origin, life-origin, and chemicals-origin experts. It is where scientists and laypeople get all their origin-related questions properly answered. It is the only place where the accurate interpretation of the universe-origin, life-origin, and chemicals-origin data matters a lot.

For Nathanael-Israel, decoding the origin of the universe and everything in it is not a job, but his life mission, and helping others to fully understand that is his mission. Visit Science180Academy.com today to start.

Science180's clients and prospects have a profound technical knowledge and background in science, while others don't. Some are creationists, others are anti-creationists. Some are believers, others are freethinkers (including atheists, humanists, rationalists, agnostics, nontheists, nonreligious, skeptics, nonbelievers, religiously unaffiliated, spiritual-not-religious, ex-believers, and doubters).

Regardless of their background, belief, or disbelief, Science180 works with each of these people to figure out their needs, priorities, and the products and services that best fit them. Science180 improves their knowledge, experience, performance, and answer their questions (related to the universe-origin, life-origin, and chemicals-origin) by crafting a personalized program that perfectly matches their interests, needs, and things that are dear and meaningful to them whether it is to:

- Protect yourself and loved ones by keeping all of you secured and empowered with the true knowledge of the origin of the universe
- Have a reliable access to the world's authority on origin-related matters and get your origin questions professionally answered with the truth step-by-step

Science180: Accurately Understand Universe-Origin and Life-Origin. Be Happy Forever!

- Connect with practical tips about how to decode the origin of the universe, life, and chemicals and protect yourself from wrong theories in the literature and the media
- Ultimately boost your confidence in detecting, confronting, and avoiding wrong theories by knowing the facts and processes involved in the formation of the universe
- Enjoy multiple origin-related programs and choose the ones that best suit your needs
- Benefit from continual updates and assistance during your journey to decode the universe, and clear your way for the universe-origin related freedom, power, technology, innovation, and breakthroughs of the future.
- Scientifically test and know whether there is a God that created the universe or not, and which God it is
- Free yourself from boring explanations of the origin of the universe, life, and chemicals and embrace the proven theory that opens doors to unparallel opportunities
- Disrupt all religious and scientific chains of repetitive nonsenses about the universe-origin, life-origin, and turn your attention toward unconventional ideas leading to greater innovation and prosperity
- Satisfy your burning desire for freedom from beliefs and scientific theories about the universe-origin and life-origin that suffocate you and bind your mind, faith, unbelief, heart, and education
- Stand as the lightning bolt that electrifies your friends who are still struggling to understand the universe-origin
- Fearlessly push the boundaries of the human abilities to properly understand what is perceived as un-understandable, mysterious, supernatural, unimaginable, impossible, and unthinkable that holds you back
- Empower and align yourself with Science180, the historic company that has done what no other organization has ever done: accurately decode the origin of the cosmos and its content

To register or to learn more, visit Science180Academy.com today.

Owned by Science180, Science180 Academy is a training, speaking, consulting, and mentoring program specialized in everything universe-origin, life-origin, chemicals-origin, and anything at the intersection of reason and faith, or science and religion.

Science180 Academy deals with different subjects according to the needs of its members or target groups. When people register to Science180 Academy, they must choose the program(s) they want to focus on so their training can be properly personalized accordingly. This is similar to how people register to a university, and take classes in a specific department matching their needs!

Science180's breakthroughs are so complex and dense that it is not realistic or good to try to explain all in just one academy, else people will be overwhelmed, disinterested, and confused by the plethora of data to handle. In other words, Science180 Academy offers a wide range of origin-related training in various domains strategically designed to allow people to choose the most suitable for their needs so that, regardless of their background or field of expertise, people can equip themselves, align their mindset, improve lives today and forever using the accurate explanation of the origin of the universe, of life, and of chemicals. Science180 Academy curriculum is based on 12 years of deep unconventional research that culminated with the publication of many much-admired books on the formation of the universe and its content:

- "Turbulent Origin of the Universe"
- "Reconciling Science and Creation Accurately"
- "Turbulent Origin of Chemical Particles"
- "Origin of the Spiritual World"
- "From Science to Bible's Conclusions"
- "Turbulent Origin of Life"
- "How Baby Universe Was Born"
- "Science180 Accurate Scientific Proof of God"

The content of each Science180 Academy is strategically crafted by Dr. Nathanael-Israel Israel (who is acknowledged as the internationally-acclaimed world's authority in origin-related issues) to suit both scientists and nonscientists, religious and nonreligious people, leaders as well as followers, so they can fully decode the proofs of the formation of the universe, of life, and of chemicals they have been wanting to demonstrate or grasp.

The current programs of Science180 Academy are:

1. **SCIENCE180 ACADEMY OF COSMOLOGY** (Designed for all scientists who want to scientifically study cosmology, the science of the origin and fate of the universe)

2. **SCIENCE180 ACADEMY OF TURBULENCE** (This is a perfect fit for scientists and other experts interested in studying abiotic turbulence). Examples of these people include:

- o   Automobile manufacturers
- o   Chemical producers
- o   Commercial space businesses / aerospace manufacturers
- o   Private and governmental space agencies
- o   Turbulence companies, experts, and scientists

3. SCIENCE180 ACADEMY OF LIFE SCIENCES (Tailored to those who want to study biotic turbulence):
- o   Life science specialists, associations, or organizations
- o   Healthcare companies
- o   Biotechnological companies, etc.

4. SCIENCE180 ACADEMY OF CHEMISTRY (Designed for chemists, biochemists, scientists, and other educated people who want to understand the origin of chemical particles)

5. SCIENCE180 ACADEMY FOR LAYPEOPLE OR THE GENERAL PUBLIC (Very fit for any layperson or "less" educated people who wants to learn (in a simple language) deep insights that even those who went to university for years were unable to decrypt by themselves, so these laypeople can be equipped to eliminate all forms of scientific and religious universe-origin prejudices)

6. SCIENCE180 ACADEMY FOR CHILDREN (This Academy breaks down origin key topics into language that children can fully understand). This is the only Science180 Academy that your whole family will like and enjoy together, and which will set children on the path of success by accurately showing them early in life the formation of the universe, and how to detect errors in theories or stories that would misguide them as they grow up.
Therefore, you need to add this great, efficient, trustworthy, and cost-effective "Science180 Academy for Children" to the strategic journey of children toward their best tomorrow.

7. SCIENCE180 ACADEMY OF CREATIONISM (Science180 Creationism is a scientific theory spearheaded by the groundbreaking discoveries of Nathanael-Israel Israel, that scientifically explained the origin of the universe, life, and chemicals using turbulence, and that mathematically reconciled science and the Biblical account of creation for the first time in history. Science180 is different from all existing creationist theories known before 2025. Science180 Creationism reconciled science with the Biblical account of creation, including scientifically proving that the Earth was formed on Day 3, while the Moon and the Sun were formed on Day 4 of creation!). As you attend "Science180 Academy of Creationism", you will receive accurate answers to all your questions concerning the creation of the universe). The target audience of "Science180 Academy of Creationism" include:

- o Christians
- o Christian or Bible colleges, universities, and schools
- o Biblical creation experts
- o Churches, Christian ministries, televangelists, pastors, prophets, teachers, apostles, and all other ministers of God
- o Christian associations
- o Anti-creationists wanting to explore the Biblical creation narrative

8. SCIENCE180 ACADEMY OF THE PSEUDEPIGRAPHA AND SPIRITUAL WORLD (Only one ancient blueprint has the reliable power to help you to accurately decrypt the spiritual origin and history of everything in the universe. If you are a believer and want to delve into the prophetic, angelic, and higher order of knowledge based on the spiritual world, then this Science180 Academy is for you. This program is suitable for those who took at least "Science180 Academy of Creationism". For you to enjoy the courses in this Academy, you need to have at least learned about or attended Science180 Academy of Creationism. If not, you may waste your time trying to grasp simple and supernatural things that cannot be scientifically proven in this Academy, but in Science180 Academy of Creationism).

9. SCIENCE180 ACADEMY FOR FREETHINKERS & ALL ANTI-CREATIONISTS (This Science180 Academy is designed for evolutionists, anti-creationists, and all other types of unbelievers seeking to rationally explore and understand alternative arguments for creation or formation or origin of the universe, life, and chemicals from a fresh, scientific perspective).

10. SCIENCE180 ACADEMY OF LEADERSHIP-(Also called "Science180 Academy for Leaders", this program will enlighten leaders of organizations on how to solve their people problems, process problems, and profit problems related to the origin of the universe, of life, and chemicals according to their domain of expertise). With "Science180 Academy of Leadership", leaders will gain new insights so they can cast new visions and avoid focusing on screwed-up processes, products, and services related to universe-origin initiatives that need to be fixed, faced, or dealt with.
Science180 Academy of Leadership will also equip leaders to address process problems related to inefficiency, gaps, missed opportunities, wasted time and efforts, useless layers between organization and customers concerning the innovation, research methodology, strategic planning, … in alignment with the historic Science180 breakthroughs so that they can sell more often at full price, avoid regrets in the end, open new markets focusing on real solutions,

stop wasting time on useless products that will yield nothing, start focusing on the real money-making problems, … and outpace their competition as long as their products or services are related to the origin of the universe, life, and chemicals. Perfect fits for Science180 Academy of Leadership include leaders of:

- o Scientific organizations, Academies of sciences
- o Universities, colleges, and schools
- o Automobile manufacturers, Governmental and private space agencies, Technology companies
- o Chemical producers, Healthcare and biotechnology companies
- o Any other organization that can benefit from the insight into the origin of the universe, life, and chemicals

11. SCIENCE180 ACADEMY FOR GOVERNMENTAL AGENCIES (Do you want to know how and why most nations and governments are wasting millions of dollars on universe-origin and life-origin researches they don't need … and how to avoid it? Indeed, for most developed nations, and even for some under developed countries, universe-origin projects can cost billions of US dollars and other expensive things that cannot be afforded without sacrificing crucial priorities. Even in developed countries, the impact and the return of investment of the space researches are subject of intense political and economic debates. What if your nation or institution can reduce wasteful spending on universe-origin research and life-origin research, as well as your dependency of wrong theories on the origin of the universe and life?

"Science180 Academy for governmental agencies" will show you how to use the latest scientific breakthrough to better understand the origin of the universe without wasting money on what is already known or what we think we don't know, but that most scientists ignore. Dr. Nathanael-Israel Israel delivers science-backed insight to properly understand all the processes connected to the universe formation—so you don't waste more money and time on trying to research the beginning of the cosmos, but to focus on reducing budget of spatial agencies, focus on real science, cutting-edge research, and things that inevitably lead to discovery and innovation). Perfect matches for "Science180 Academy for governmental agencies" include leaders of:

- o Department of Defense and Department of Energy
- o Aeronautics / Space agencies, National Science Foundation,

12. OTHER SCIENCE180 ACADEMY: If you did not relate with any of the Science180 Academies mentioned above, but you are still interested in learning something specific about the origin of the universe, life, and chemicals that better fits your needs, please visit Science180Academy.com to contact us.

'Science180 Academy' Success Strategy:
SCIENCE180 SERVICES AND PRODUCTS YOU WILL LOVE

Because you are reading this book, you are probably very interested in answering your questions about the origin of the universe, of life, and of chemicals. Imagine you want to be trained by Dr. Nathanael-Israel Israel and his team so you can benefit from their outstanding expertise to empower yourself or your team. Or you want him to give a keynote speech, a seminar, or any other kind of talk or conference at your organization. Or you want him to mentor you or some people or team at your organization. Maybe you have critical origin-related questions that you need his help to accurately answer. You want a true expert to talk with you about the customized program or game plan that fits your needs. You want him to tailor his advice, expert feedback, and proven shortcuts to the stage of life you are in and help you get to where you want to be in your desire to properly understand the origin of the universe, life, and chemicals and harness the benefits that come with it. Perhaps you don't know how to properly get any of these important tasks done according to your specific needs or the needs and demands of your organization. That is what Science180 Academy is all about. Visit Science180.com/services for more details about how to benefit from the services that Science180 provides.

Maybe you are a leader that wants to hire Dr. Nathanael-Israel Israel and his team to train some departments at your organization. Or you want to refer them to other companies like a good dish passed around the dinner table, and you want to explore how Nathanael-Israel Israel can pay you something for that referral. Maybe you attended Nathanael-Israel Israel's speaking program, for which, without going into details, he accurately raised your awareness about how the universe, life, and chemicals were formed. Or maybe you attended his training, in which he detailed and showed you how he decoded the scientific data using various tools and certain thinking strategies that helped him and which transferred some skills to you; and now, you are interested in a long term one-on-one consulting, or mentoring program with him, so that, he delves into more details about how to use proven techniques to decode the universe (strategies for data collection, data analysis, data presentation, writing, and even tips for future research) and change your behavior on a long term basis. If you related to any of the points mentioned above, Science180 Academy is the right fit for you!

Other customizable services that Science180 provides include:
- Books and other products (e.g. booklets, online courses, posters, how-to-guides, study guides, and field guides)
- Book publishing (Yes! Science180 can publish your books!)
- Conferences

Science180: Accurately Understand Universe-Origin and Life-Origin. Be Happy Forever!

- Consulting
- Podcasting
- Retreats
- Seminars
- Speaking engagements (offline and onsite–e.g. seminar, keynotes, and workshops)
- Training
- Video programs
- Virtual presentations

Here are other reasons why you should choose to work with or hire Nathanael-Israel Israel and the team at Science180:

- A simple universe-origin and life-origin theory that made no assumption
- Accurate universe-origin and life-origin decoding trailblazer
- Bringing people together through the power of the accurate decoding and understanding of the universe-origin and life-origin
- Complex universe-origin and life-origin questions solved accurately in a simple language
- Current hot cosmological and life-origin solutions for kids
- Discover the key variables needed to decode the universe-origin and life-origin
- Easily understand complex universe-origin and life-origin equations in minutes
- Efficient, trustworthy, and cost-effective lessons to add to your strategic journey toward your best tomorrow
- Improve your understanding of the universe-origin and life-origin with new, accurate products and services
- Light in the heart of science the lamp of understanding
- Mover of the needle on the universe-origin and life-origin vs. creationism debate
- One-stop platform for the origin of the universe, life, and chemicals
- Redefine the origin of the universe and of life accurately
- Source of unconventional wisdom and knowledge on the origin of the universe, life, and chemicals
- State-of-the-art decoding experience of the universe origin & life origin
- The best universe-origin and life-origin decoding. Only on Science180
- The go-to source for valuable universe-origin and life-origin information

Nathanael-Israel Israel: Known as the Outside-of-the-box Universe-Origin Scientist

- The most accurate, reliable, safest, best explanation of the universe-origin and life-origin ever
- The only one formula accurate enough to explain the formation of the universe and life
- The place where children get all their universe-origin and life-origin questions properly answered
- The place where science accurately tests Biblical creation
- The place where the accurate interpretation of universe-origin and life-origin data matters
- Science180: The premier organization that scientifically decoded the origin of the universe, life, and chemicals accurately
- The science that reunites your reason, faith, or doubt
- The undeniable scientific challenge to all metaphorical, figurative, loose, liberal, or vague explanations of the universe-origin and life-origin
- The unquestionable scientific challenge to wrong universe-origin and life-origin
- Theory that helps you avoid dangerous dogma and irrational thinking
- Universe-origin and life-origin formula accurately made easy
- Universe-origin and life-origin questions last bus stop
- Universe-origin and life-origin theory that helps you fight wasteful programs

Science180: Accurately Understand Universe-Origin and Life-Origin. Be Happy Forever!

'Science180 Academy' Success Strategy
SCIENCE180 SEMINARS

People whose awareness is raised by Science180 usually ask me to go deeper or they wonder "what's else?". That is one of the reasons Science180 trains them through strategic work sessions (during seminars or training sessions) that transfer customizable skills and solutions to them. Science180 Seminars are client-centered and tailored to strongly engage the clients so they maximize the discovery of and the tapping into new opportunities, and exponentially outperform their expectations. Science180 offers customizable seminars that can be labeled as a colloquy, conference, consultation, discussion, forum, keynote speech, lecture, lesson, meeting, symposium, summit, study group, tutorial, workshop or working section accordingly on any topic related to:

- Universe-origin for scientists and mathematicians, philosophers, laypeople, and the general public
- Universe-origin or universe creation for believers
- Life-origin for life scientists, for all other scientists, and for believers
- Chemical-origin for scientists
- Universe-origin seminars for children
- Universe and life-origin for pseudepigraphic believers

As you contact us with your needs, we can customize your program accordingly. Learn more at Science180Seminars.com.

'Science180 Academy' Success Strategy
SCIENCE180 CONSULTING

Because Science180's trainings, seminars, or strategic work sessions (through which it transfers skills and training solutions) are great, some customers want to go even deeper on a long-term, sustainable basis. That is where Science180 Consulting, one-on-one consulting, and mentoring (that some people may prefer calling coaching programs) come in. That is where Science180 can truly change people's behavior on a long-term basis according to their specific needs. With Science180 Consulting, you will discover and understand the deep secrets of the formation of the universe, life, and chemicals around you. Hear Dr. Nathanael-Israel Israel's personal selection and teaching on key topics that will help you break the code of the universe's formation and functioning. All strategically designed to enlighten you, guide you to navigate and filter the massive data collected on the universe and its content so you know how to answer the world's most challenging origin questions, remove any scientific and philosophical cataracts that may be blocking you, and help bring you many steps closer to your best life today and forever. Science180 Consulting will train you, transfer unconventional skills to you, and change your behavior so you go deeper. To get started today or to learn more, go to Science180Consulting.com.

'Science180 Academy' Success Strategy
SCIENCE180 MASTER CLASS

Hear the greatest scientific and philosophic lessons from top scientists, philosophers, thinkers, and public figures who have realized historic mistakes they made in life (concerning the origin of the universe, life, and chemicals), and that they corrected thanks to the historic discovery of Nathanael-Israel Israel, the world's first 180Scientist who founded Science180 and who is known as the one who truly decrypted the universe origin for the first time. In their own words, these renowned personalities share with the world key lessons they have learned in life and how people can learn from their experiences to improve lives instead of repeating their mistakes that many people still ignore at their own perils. To learn more, contact us at Science180.com/contact.

Science180: Accurately Understand Universe-Origin and Life-Origin. Be Happy Forever!

'Science180 Academy' Success Strategy:
SCIENCE180 MODELS OF THE ORIGIN OF THE UNIVERSE AND ITS CONTENT

Science180 Models consist of all the theories elaborated by Nathanael-Israel Israel regarding his ground breaking discovery on the origin of the universe and its content including all forms of life and chemical particles. These theories are detailed in various books written by Dr. Nathanael-Israel Israel encompass the following:

1. SCIENCE180 MODEL OF COSMOLOGY, also called Science180 Cosmology, Science180 Model of Cosmology, Science180 Cosmological Model, a scientific theory that explains Science180 to the scientists. Discover the details of this model in Nathanael-Israel Israel's book titled *"Turbulent Origin of the Universe"*. In that book, you will also unearth the new physics that will revolutionize science forever and land you into a zone of original ideas that improve lives nonstop regardless of your expertise.

2. SCIENCE180 CREATIONISM, also called Science180 Model of the Creation of the Universe and Life by God, a scientific theory that presents the origin of the universe in a biblical language. If you want to learn more about how to scientifically prove the Biblical account of the creation of the universe and the existence of God in a way that makes the head of God deniers to spin faster than a DJ's turntable, then get Nathanael-Israel Israel's book titled *"Reconciling Science and Creation Accurately"*.

3. SCIENCE180 MODEL OF THE ORIGIN OF CHEMICAL PARTICLES, a scientific theory that explains the origin of chemical particles with the perspective of Science180 Turbulence. If you want to professionally learn how to transform the true knowledge of the origin of chemical particles into insights that significantly add value to your life in less time, successfully establish you as a symbol of freedom, power, creativity, and originality in your field of expertise, get Nathanael-Israel Israel's book *"Turbulent Origin of Chemical Particles"*, THE ultimate how-to guide for great people wanting to correctly decode the origin of the chemicals and positively transform their lives. Get this celebrated book today. Don't wait!

4. SCIENCE180 MODEL FOR THE GENERAL PUBLIC (which explains the origin of the universe and life to the general public in a language that laypeople can understand). Find out more in Nathanael-Israel Israel's book called *"From Science to Bible's Conclusions"*, a scientifically verifiable, bestselling book to finally get the accurate, jaw-dropping answer that has been rationally shaking believers, skeptics, and freethinkers. Get this very popular book today.

5. SCIENCE180 MODEL OF LIFE-ORIGIN, or Science180 Model of the Origin of Life, a scientific theory that explains the origin of all forms of life using turbulence. To unlock the step-by-step pathway to decode the origin of life and get the power, freedom, and boldness to detect, correct, and remove all misinformation, ambiguity, and misleading claims and theories surrounding the life-origin and take advantage of the opportunities that an accurate understanding of the life-origin creates, get Nathanael-Israel Israel's book titled *"Turbulent Origin of Life"*.

6. SCIENCE180 MODEL FOR CHILDREN, a children's version of the theory of the origin of the universe and life in a language that 7-12 years old children can properly understand. To know the proven formula that helps children to easily answer their huge universe-origin and life-origin questions with confidence, humor, and joy, get *"How Baby Universe Was Born"*, the pragmatic book that has been causing children to belly laugh and thank those who offered it to them.

7. SCIENCE180 MODEL OF PSEUDEPIGRAPHA, a deep explanation of the secrets of the origin of the universe and life revealed a long time ago, but hidden from the general public. To discover how the only one ancient blueprint has the reliable power to help you to accurately decrypt the spiritual origin and history of everything in the universe, get Nathanael-Israel Israel's book called *"Origin of the Spiritual World"*. In it, you will discover deep rejected secrets that have prevented humankind from unearthing the beginning of the universe and know how to properly use the lost and rejected scriptures to articulate the process by which the universe was formed, so you can use that insight to improve your understanding of the Bible, innovate in your domain of interest, and improve your life perpetually.

8. SCIENCE180 MODEL OF THE PROOF OF THE EXISTENCE OF GOD, a theory that ties together most of Nathanael-Israel Israel's discoveries that scientifically prove the existence of God. With Nathanael-Israel Israel's book *"Science180 Accurate Scientific Proof of God"*, you will surely know the only way to scientifically know if God exist, and if so, which of the thousands of beings worshipped across the globe is the true God. In that book, you will also discover the errors in the scientific and religious theories (about the origin of the universe, life, and chemicals) that are putting you at a high risk you will never recover from if you don't quickly and confidently learn how to rationally take control over threats lurking at the edge of your efforts to understand the universe and life today.

9. SCIENCE180 THEORY OF EVERYTHING, (also called the theory of all theories), ties together everything in the universe into a single theory. Checkout Science180.com to learn more about the incoming book that covers this extremely important topic.

26. NEXT STEPS OF THE JOURNEY

Get free resources on Science180.com

If you have finished reading this book and would like to learn more about my discoveries and how they can help you, you are at the right place. Indeed, I am committed to helping you address any questions you may still have about the origin, function, and fate of the universe, and how you can partner with me to achieve greater results.

To get free resources that will help you understand other aspects of the universe formation not covered in this book, visit Science180.com and my personal website, Israel120.com. On those sites, I will be sharing guides and strategies to get the most out of my initiatives. I will also be sharing my favorite references, tips, next-steps readings, and other important things in the pipeline to help you, regardless of your field of expertise, interests, or needs.

Subscribe to "Science180 Newsletter": The only accurate universe-origin, life-origin, and chemicals-origin newsletter in the whole world!

Be a part of decoding the universe-origin, life-origin, and chemicals-origin! Get origin-related news, information, discoveries, updates, announcements, reviews, articles, educational materials, and opportunities, from a holistic perspective not available anywhere else, so you can participate in and enjoy decoding the origin, current state, and fate of the universe and its content. You will also receive priceless tips about how Nathanael-Israel thinks, what his secrets and initiatives are, what he has accomplished, and what he recommends. Without any delay, sign up for the Science180 Newsletter today at Science180.com/newsletter. It is free!

Speaking engagement

In addition to writing groundbreaking books and pursuing other business endeavors, Nathanael-Israel Israel is a renowned speaker you can invite to speak at your organization.

Values that Dr. Nathanael-Israel Israel can add to your life include:
- Rare expertise and tips that will increase your abilities

- Usefulness that will advance your impact regardless of your field of expertise
- Understanding of the world that will sharpen your perspective
- Critical information that will positively change your life
- Experiences turned into insight that will motivate and guide you
- Irrefutable scientific proofs of the existence of God that will save you time and launch you into a zone of unlimited opportunities
- Unquestionable scientific proofs of how God created the universe
- Accurate demonstration of the historic formula that reconciled science and the Bible
- Enlightenment that will help people, including Christians, to start using their brains instead of just praying and expecting God to do everything for them

For speaking inquiries, including how to book Dr. Nathanael-Israel Israel to speak to your organization or at an event, visit Science180.com/speaking for more details.

As the standout scientific authority who accurately decoded the universe, Nathanael-Israel Israel has been helping countless people across the globe discover and understand the universe's complex origins without omitting the challenging questions that people of all ages have struggled to answer for thousands of years! As the true go-to expert on the formation of the universe and life, Nathanael-Israel believes that, regardless of age, background, culture, religion, or profession, everyone deserves to understand how the universe and life formed and how to leverage that knowledge to improve their lives nonstop. Therefore, his groundbreaking discoveries about the formation of the universe, life, and chemicals have been distilled into books tailored for scientists (including physicists, chemists, biologists, mathematicians), laypeople or the general public, believers and freethinkers, philosophers, children, etc., thereby maximizing the benefits to humanity.

When you hire Nathanael-Israel Israel to speak at your organization, you will:

- Get specific in-depth knowledge, up-to-the-minute information, ideas, and insights about the universe-origin, life-origin, and chemicals-origin so that you expand your market, cut useless costs, stop wasting time on inadequate projects, and start focusing on the profitable solutions
- Get relevant universe-origin stories that are specific to your field of expertise
- Learn from a cooperative, flexible, and an easy to work with expert who will respond to your universe formation needs and position you to stay on top of your competitors

- Interact with a renowned expert who will not just lecture you, but will help you sort out your origin-related questions using strategies to tap into deep secrets you ignore

- Listen to an experienced expert who discovered outstanding secrets about the origin of all there is

- Learn authentic information not from someone who reads you a PowerPoint, but from the true go-to expert (when it comes to critical cosmological problems) who will share with you both his mistakes and successes that will help you get much closer to the better life you want to live

- Revolutionize every origin-related domain with your accurate understanding of the universe-origin

- Scientifically learn how the Earth was formed on the 3rd day of creation

- Logically learn how the Sun and the Moon were formed on the 4th day of creation

- Hear Dr. Nathanael-Israel Israel's personal selection and teaching of key topics that will help you break the code of the universe formation and functioning, and strategically enlighten you, guide you to navigate and filter the massive data collected on the universe and its content so you know how to answer the world's most challenging origin questions, remove any scientific and philosophical cataracts that may be blocking you, and help bring you many steps closer to your best life today and forever

- Hear the greatest scientific and philosophic lessons of some top scientists, philosophers, thinkers, and public figures who have realized historic mistakes they made in life (concerning the origin of the universe, life, and chemicals), and that they corrected thanks to the discoveries of Nathanael-Israel Israel, who founded Science180, and who is acknowledged as the scientist that truly decrypted the universe-origin for the first time

- Get world key lessons successful people have learned in life, and how people can learn from their experiences to improve their lives instead of repeating their mistakes that many people still ignore at their own perils

To book Dr. Nathanael-Israel Israel for a speaking engagement, visit Science180.com/speaking.

How the adult in your home can make money by joining the affiliate program to sell Nathanael-Israel Israel's books
Greetings,
Do you want to make easy money by selling the #1 universe-origin, life-origin, and chemicals-origin books on your website, newsletter, and by mail? You can start making big money by helping sell Science180 Books, including this one, on your website and through your network. Indeed, by now you know that I operate a

website called Science180.com, which specializes in helping people around the globe scientifically decode and understand the formation of the universe, life, and chemicals.

Your contacts, site, blog, forum, podcast, and newsletter may be admired among my target audience. Some of my products and services may be of interest to your audience. My books are the first in history to scientifically demonstrate the match between science and Biblical creation in a way that satisfies both believers and nonbelievers, a historic achievement and discovery that is revolutionizing our view of the origin of the universe, life, and chemicals for the benefit of humankind.

Imagine you have a website where you can talk to people about my books and services, and get a great percentage of every purchase they make on my site? Imagine you send a link to my books to your friends or network, and when any of your contacts buy a copy, you get a percentage of what they pay on my sites. Imagine you can email your friends to spread the good news about my books, and when anyone uses that link to buy them, I give you something. Well! This is what the affiliate program is about. Apply today or learn more about it at Science180.com/affiliate. Likewise, if you own a website, you can apply for Science180's affiliate program, and I will send you a specific affiliate link to place on your website and in your newsletter. If people click on it to buy my books, they will be taken to my page, and after they buy, I will pay you a set amount, sharing the profit with you instead of just saying thank you.

Would you be interested in reviewing some of my products and services to explore becoming an affiliate? We have a wonderful affiliate program, and commissions are paid quickly and accurately.

If you are satisfied with the quality of our products and services, I am convinced you will also be impressed by our affiliate program.
I look forward to hearing from you

Nathanael-Israel Israel, PhD

Collaborate or partner with Nathanael-Israel Israel
If you have any lawful idea, initiative, or suggestion for a genuine partnership with Dr. Nathanael-Israel Israel or Science180, please visit Science180.com/partner to inform us.

How to be trained or mentored by or have a one-on-one consultation with Dr. Nathanael-Israel Israel
Hire Nathanael-Israel Israel to train you or your organization in the best ways to conduct yourself and your organization to align your initiatives with the real understanding of the origin of the universe, of life, and of chemical particles in a way that you will not hear anywhere else. Nathanael-Israel Israel offers training

through the "Science180 Academy" program. For training purposes, please visit Science180Academy.com.

Visit Nathanael-Israel Israel's personal website to get for free great resources you won't find anywhere else

To stay in touch with Dr. Nathanael-Israel Israel, and to get updates directly from him, please visit his website, Israel120.com, and sign up for his popular newsletter at Israel120.com/newsletter for free.

Ask for review

If you are a book reviewer or a professional wanting to review this book or others written by Nathanael-Israel Israel, please get in touch with us at Science180.com/AskForReview

Donate and support Nathanael-Israel Israel's efforts and initiatives

To help humankind accurately understand the real origin of the universe and its content, as I have done in the groundbreaking books I published after 12 years of sacrifice. If you would like to contribute, please visit Israel120.com/donate or Science180.com/donate. Your donation will help me continue doing what I did to bring these books to life, which you enjoyed and know will help many people across the globe. No amount of money is too small or too big. Whatever you can give, please give.

Quantity discounts: Purchase Science180 books, including this one, in bulk at a special discount

To purchase Science180 books, including this one, in bulk at a special discount for sales promotion, corporate gifts, fund-raising, or educational purposes, or to create special editions to specifications, visit Science180.com/discount.

Buy a copy of Nathanael-Israel Israel's books for your friends, family, or someone else

If this book has been a blessing to you, and we know it has, please consider getting another copy and giving it to a friend, a family member, or someone you think it may help or challenge. If you want many copies, we can offer you a discount; just contact us as we previously explained.

Recommend Nathanael-Israel Israel's books to your organization

Because I know this book has been a blessing to you, I ask that you recommend it, along with others I wrote, to your organization, class, workplace, church, school, network, or clubs. Recommending this book will help others tap into the blessings and opportunities that my books will open for them.

Share Nathanael-Israel Israel's groundbreaking discovery with others

To reach more people, please share Nathanael-Israel Israel's findings with others, as many people still do not understand how the universe was formed, and sharing your experience of reading this book will help them. If you enjoy Nathanael-Israel Israel's books, please help other people find them by writing a book review on your blog or on online bookstores, or write it and share it with us. Likewise, share and mention this book on your social media platforms (e.g., Facebook, Twitter, YouTube, etc.).

Follow Nathanael-Israel Israel on social media

In our modern world, social media has become a major factor in how messages spread across the globe. To ensure more people hear about the good news revealed in my books, I need you to follow me and share my content on your social media and in your network. To know the full list of my social media accounts and follow me, please visit Science180.com/socialmedia.

Share your feedback, criticism, testimony, experience, adventures, story, or comment about this book with me

How have Nathanael-Israel Israel's books and services at Science180 improved your life? I would love to hear from you.

To better help you next and encourage others, I need to hear and capture your testimony or criticism. Please visit the feedback page, Science180.com/feedback, to tell me:

- How this book impacted you or will impact you
- What you like or dislike or disagree with
- What do you think, wish, or dream that I need to work on next
- What you wish to see in this book, but that was absent
- What shocked you the most
- What got your heart pumping as you were reading this book
- What did you find more insightful or thought-provoking
- What do you want to do to be a part of my journey
- How has my work changed your life or someone else's life

Message from the publisher of this book

Just like Nathanael-Israel Israel, you can publish your book(s) with us, too. To get started and see how we may help you, please visit Science180Publishing.com today.

To contact Nathanael-Israel Israel or Science180

For any suggestions or questions, please visit Science180.com/contact and Nathanael-Israel Israel's personal website: Israel120.com. Feel free to ask me any questions you have about the formation of the universe, life, and chemicals.

Nathanael-Israel Israel: Who Happens to be the World's #1 Authority on the
Turbulent Origin of the Universe and Life

27. REFERENCES

NASA (2018). Planetary fact sheets. Fact sheets of the Sun, planets, satellites, rings and selected asteroids in the Solar System. Author/Curator: Dr. David R. Williams, NASA Goddard Space Flight Center, Greenbelt, MD, USA. http://nssdc.gsfc.nasa.gov/planetary/factsheet/. Visited on November 19, 2018.

Israel Nathanael-Israel (2025a). Turbulent Origin of the Universe. Science180, Augusta, USA 683 pages.

Israel Nathanael-Israel (2025b). From Science to Bible's Conclusions. Science180, Augusta, USA 170 pages.

Israel Nathanael-Israel (2025c). Reconciling Science and Creation Accurately. Science180, Augusta, USA 299 pages.

Israel Nathanael-Israel (2025d). Turbulent Origin of Chemical Particles. Science180, Augusta, USA 397 pages.

Israel Nathanael-Israel (2025e). Turbulent Origin of Life. Science180, Augusta, USA 370 pages.

Israel Nathanael-Israel (2025f). Origin of the Spiritual World. Science180, Augusta, USA 151 pages.

Israel Nathanael-Israel (2025g). How Baby Universe Was Born. Science180, Augusta, USA 130 pages.

Israel Nathanael-Israel (2025h). How God Created Baby Universe. Science180, Augusta, USA 224 pages.

Israel Nathanael-Israel (2025i). Science180 Accurate Scientific Proof of God. Science180, Augusta, USA 214 pages.

28. INDEX

INDEX

Nathanael-Israel Israel: The Scientific Defender of God's Existence

ABOUT THE AUTHOR

Dr. Nathanael-Israel Israel is the founder of Science180, the American company which mission is to improve the current and future state of human beings by accurately decoding and teaching them the real origin and formation of the universe, of life, and of chemicals, and meaningfully engaging business, nonprofit, political, academic, civil society leaders and followers to properly shape local and global agendas that authentically value the truth. As the creator of the Universe Turbulent Origin Formula™. Dr. Nathanael-Israel Israel has revolutionized the way billions of people around the world think about the origin of the universe, of life, and of chemicals. Nobody understands and teaches the formation of everything in the universe (e.g., the Milky Way Galaxy, the Sun, the Earth, the Moon, and all other galaxies, stars, planets, satellites, and asteroids) better than Nathanael-Israel Israel. Individuals and organizations across the globe have been calling him so that he helps them scientifically unlock the code of the universe-formation, helping veterans and rookies to have the real keys to decrypt the universe and turbulence (one of the top biggest mysteries in science) from the historic, unique, accurate, simple, easy-to-understand, nonconformist, trailblazing perspective that anybody can quickly learn at Science180 Academy (Science180Academy.com).

Science180 Academy delivers outstanding value, insight, and lessons to help people accurately understand the true origins of the universe, chemistry, and life, so they can tap into that knowledge to improve their lives over time. Nathanael-Israel's goal is to give you practicable and undeniable proofs of the formation of the universe so you can be fired up to become the best version of you, and to cause positive changes to your initiatives that will profit you today and forever. For Nathanael-Israel, accurately decoding and teaching the origin of the universe and everything in it is not a job but his life's mission, and helping others fully understand it brings him closer to his assignment. He is also a father who faced the same challenges as most parents in teaching his children about the formation of the universe.

Dr. Israel earned his PhD in Plant, Insect, and Microbial Sciences in the USA, where he graduated first in his class of hundreds of PhD candidates. This Beninese-American is a member of the American Chemical Society, American Association for the Advancement of Science, American Society of Agricultural and Biological Engineers, American Society for Microbiology, American Society of Biochemistry and Molecular Biology, Ecological Society of America, American Society of Agronomy, Crop Science Society of America, and Soil Science Society of America. A scientist, a mathematician, a consultant, and the owner of Global Diaspora News, a news company in the USA, Dr. Israel is the author of the popular books:

- Turbulent Origin of Chemical Particles
- Turbulent Origin of Life
- From Science to Bible's Conclusions
- How Baby Universe Was Born
- How God Created Baby Universe
- Science180 Accurate Scientific Proof of God
- Turbulent Origin of the Universe
- Reconciling Science and Creation Accurately
- Origin of the Spiritual World
- Mathematical Proof of God's Existence at the Intersection of Science and Faith.
- Boldest Scientific Formula of God and Creation

If you want to accurately understand the origin of anything, be sure to get a copy of these amazing books. You cannot afford to ignore the greater, better, faster, simpler, cheaper, easier, and accurate formulas unlocked in these important books that successfully cracked the origin of the universe, of life, and of chemicals in a language that scientists, laypeople, adults, children, believers, skeptics, and anybody else can properly understand and enjoy.

Visit Israel120.com today to connect with this historic discoverer of the all-in-one proven and uncomplicated formula that accurately decoded the origin of the universe, of life, and of chemicals.

Nathanael-Israel Israel: The Scientific Defender of God's Existence

TAKE A NOTE

Nathanael-Israel Israel: The Scientific Decoder of the Universe's Origin.
Learn more at www.Science180.com